A Field Guide to the Rattans of Lao PDR

Tom D. Evans, Khamphone Sengdala, Oulathong V. Viengkham and Banxa Thammavong

Line drawings by Banxa Thammavong

Researched and produced with funding from the **Darwin Initiative for the Survival of Species** (operated by the Department of the Environment of the UK government) through the project 'Diversity and Sustainable Use of Rattans in Lao PDR'. This is a collaborative project between the Oxford Forestry Institute, Department of Plant Sciences, Oxford University, UK, the Royal Botanic Gardens, Kew, UK and the Forestry Research Centre, National Agriculture and Forestry Research Institute, Vientiane, Lao PDR.

Production Editor: S Dickerson

Cover and book design by Jeff Eden, page make-up by Media Resources
Information Services Department, Royal Botanic Gardens, Kew

All photography by T. Evans except p.46 and p.72, right, (J. Dransfield) and
p.80 (middle P. Boyce, bottom Media Resources)

ISBN 978 1 84246 009 2

This book is dedicated to the many herbaria around the world that have freely given loans, information and access to specimens.

Their fundamental contribution to the development of forestry and agriculture in Lao deserves our recognition and gratitude, in the hope that it will continue in the future.

Anyone interested in learning more about rattans in Lao is welcome to contact the Darwin Rattan Research Team at the Forestry Research Centre, National Agriculture and Forestry Research Institute, Ministry of Agriculture and Forestry, Vientiane. Write to: Darwin Rattan Research Team, Forestry Research Centre, Box 7174, Vientiane, Lao PDR or by email at frcfiss@laotel.com. Tom Evans can also be contacted at: Oxford Forestry Institute, South Parks Road, Oxford, OX1 3RB, UK or by email at tom.evans@care4free.net.

When more research has been done we hope to produce another edition of this book. Please send us your suggestions for additions and changes.

Printed in the UK by Marston Book Services Ltd

Contents

Sample specimen field data sheets (inside front cover)

Introductory sections

The purpose of this book

This book is intended to support the improved management of Lao rattans. Users of this book should, with practice and care, be able to identify most of the rattans they encounter. Once the scientific name of a rattan is known users can go on to learn all that has been discovered and published about that species, both in Lao and elsewhere. This is a crucial step in making the best use of rattans for the benefit of the Lao people.

Rattans contribute greatly to the Lao economy by producing flexible canes and edible shoots which are used within country or exported. They are a prime example of the value of biodiversity, the rich variety of living things which have made Lao their home. Maintaining this valuable resource is one of the reasons why Lao has recently declared many large protected areas. Yet the productivity of wild rattan populations is under threat. We believe there is scope for improved production both through cultivation (using seed from the protected areas) and better planning of harvests from the wild. There are many species of rattans in Lao and they vary greatly. For instance, some resprout after cutting, others die. The growth rates vary from a few centimetres per year to several metres. Stem quality, soil preferences and fruiting season all vary from one species to the next. Without this information, efforts to plant or manage rattans will falter. With it we hope that a more profitable and sustainable rattan sector can be developed in Lao.

Who is the book for?

This book is for people working in Lao or nearby areas who need to identify rattans for forestry, agroforestry, agriculture, conservation or scientific purposes. It is based on careful scientific research so you can use it with confidence, and it is written in a straightforward way so that people without special botanical training can use it. We are producing both Lao and English-language versions so that it can reach the widest possible audience.

How is the book arranged?

Background sections (in two parts, pp. 6–10 and 81–96):
These cover the structure of rattans and general information on their ecology and cultivation. Look here for an explanation of ideas or terms from the species accounts that you do not understand.

Field keys (pp. 10–16):
These tables guide you through the process of identifying an unfamiliar species. For initial identification start here, but remember to check the next section for confirmation.

Species accounts (pp. 17–80):
These provide enough information for you to confirm the identity of each species and give more detail on appearance, ecology and cultivation. Page 17 explains the different parts of each species account.

What are rattans?

Rattans are spiny palms with scaly fruits. Most of them have long, flexible climbing stems and long hooked whips which they use to cling onto other trees for support. This book also includes several non-climbing species which are such close relatives that they might be confused in the field. The Lao word used for rattans (*wai*) is also applied to at least three other species, two of which are not even palms, so take care! These three species are illustrated on page 80.

The scope of this book

The focus of this book is Lao People's Democratic Republic (or simply 'Lao'). But since the rattan flora of Lao is not fully known we also include information for nearby parts of other countries. These are Vietnam, Cambodia, Xishuangbanna Prefecture in Yunnan, China and the following regions of Thailand used by the Flora of Thailand Project: North, North-east, East, South-east and Central. Together Lao and these neighbouring areas are called 'Indochina' in the rest of this book.

This book includes all 50 rattan species confirmed to occur in Indochina. These are listed on page 93. They are in six genera: *Calamus*, *Daemonorops*, *Korthalsia*, *Myrialepis*, *Plectocomiopsis* and *Plectocomia*. We have recorded 32 species from Lao if we include three for which records have not yet been completely confirmed. One distinctive unnamed form, *Korthalsia* sp A from Lao, is also included but this may prove not to be a separate species. Many of the other Indochinese species in this book might be found in Lao if scientists search for them.

The book is based on a study of wild plants and herbarium specimens from Indochina. For each species we have only included information which is certain to refer to that species, because it is based on specimens. We have generally not used information from books and reports which do not mention specimens because we cannot check that the species were identified correctly.

The scientific basis for this field guide will be published separately, hopefully in the Kew Bulletin. It will include a list of all the specimens we examined and it will justify the names which have been chosen for the species in this book. During research for the book four new species of *Calamus* were discovered and named (*Calamus solitarius*, *C. oligostachys*, *C. bimaniferus* and *C. laoensis*). Four others were discovered but had not yet been formally named when this book was published (*Calamus* sp A and sp B and *Daemonorops* sp A and sp B). Many other species were found to have two or more different scientific names. To avoid confusion each species should only have one name. According to international rules the oldest name is used and other names are discarded (see the book edited by W. Greuter in the Further Reading section). Page 93 gives some alternative names which have been used in the past for the species in this book. These will be helpful if you look for information in other books using a different system of names. Further research in the future will probably cause further name changes, and may result in other new species being discovered and named.

Local names

Rural Lao people use rattans all the time, for many purposes, so of course they have names for the different species. Some people are much more skilful at remembering and assigning these names than their friends. If you find these local experts you will often find that they consistently use about as many different rattan names as a botanist would use in the same forest.

Furthermore, our informal observations suggest that they assign individuals to more or less the same species groupings as a botanist would – for example, they will probably have one consistent name for most or all of the plants that we would call *Calamus viminalis*, another name for *Daemonorops jenkinsiana* and so on. Although some more detailed studies would be useful to confirm this, it seems likely that there will be many opportunities to match up indigenous knowledge of rattans with the relevant scientific names.

However, an outsider visiting a village for the first time should be *very careful* not to hear a local name and assume he or she knows which species is referred to based on experience from another place. Glancing through the local names listed in this book will show why. Each of the local names given in this book was used by local people for a specimen we collected. Although these names are often used consistently in a particular village they vary endlessly from village to village and province to province. To pick just a few examples, *wai namleuang* (yellow spine rattan) is used for at least six botanical species, *wai hangnou* (rat-tail rattan) is used for at least four species and *wai deng* (red rattan) for at least five, from three genera! We have already recorded nine local names for *Calamus viminalis* in the Lao language, let alone names in the languages of other ethnic groups. All of these lists will surely grow as more specimens are collected. Even *wai thoon*, a name usually applied to the very distinctive and valuable *Calamus poilanei*, is used for at least two other large but commercially worthless species, *Calamus flagellum* and *Plectocomia* sp, in areas where *poilanei* apparently does not occur. In Vientiane Province the name *wai nyair* is used for both *Calamus viminalis* and *Calamus tenuis* by different people – but getting these two mixed up when you start a plantation could be a disaster – one likes dry ground, one likes flooding!

The reason for this confusing situation is simple – local names only need to distinguish the 5-15 species that might occur in the limited area used by a few neighbouring villages. It doesn't matter to these users if the same name is used for another species elsewhere. But it does matter to us, working across the whole country. If we use the wrong scientific name it will cause serious mistakes. This is easily avoided. The rule is:

Never use a list of local names from one place to find out the scientific name of a rattan in another place.

In each new area that you work you should check anew which local name or names refer to each botanical species. Do this by going out into the forest with at least two of the most experienced local people you can find and asking them the names of the rattans you see. Take good specimens of each species (see page 87) so that the scientific names can be confirmed in a herbarium. If you don't follow this process most of the information you collect will probably be of very little use to yourself or other people.

Distribution

We map the distribution of rattans in Indochina on the basis of specimens we have seen plus a few reported in published sources. The coverage for Lao, Thailand and China is quite good, but we were not able to visit herbaria in Cambodia or Vietnam, so the coverage of existing information is less complete for those countries. Most species probably occur in other parts of Indochina where they have not yet been collected.

We also summarise the distribution outside Indochina, based on published sources and specimens held at the Royal Botanic Gardens, Kew, UK. For this book, SE (South-east) China includes Guangxi, Guangdong and Hong Kong, NW (North-west) Yunnan is the Yingjiang area, NE (North-east) India is Assam plus the small states surrounding it and North-central India includes areas from Bihar westwards.

The distribution of rattans is not well known in Indochina. For example, in Lao some species are only known from one or two herbarium specimens, and only eight species are known from ten or more specimens. Only a few general comments are possible.

The rattan species present are fairly similar right across Indochina because the whole area experiences tropical monsoonal weather and there are no major barriers to prevent rattan species spreading between suitable habitats. However, rainfall, temperature and soils do differ so no one species is found in all parts of the region and each location has a slightly different community of species, even in sites at the same altitude.

The richness of species and genera seems to increase from north to south. This is perhaps due to a combination of less suitable climate and smaller areas of the preferred lowland habitats in the north. The area of forest used by a single village might contain 4-8 species in north-west Lao, 12-14 in central Lao.

The most widespread species within the region include *Calamus viminalis*, *C. tetradactylus*, *C. palustris* and *Daemonorops jenkinsiana*. Many species are widespread outside Indochina. These include some species which occur in the Thai-Malay peninsular, another group which ranges into South-east China and a third group which ranges into Myanmar and North-east India. *C. viminalis* and *C. tenuis* are both found as far away as North-central India and Java.

Twenty-two of the species in this book have never been found outside Indochina. There is no one area where these localised species are concentrated, but the most localised species appear to be concentrated in and around the Annamite Mountains (Sayphou Louang).

Species known from single sites include *Calamus harmandii*, *Calamus ceratophorus*, *Calamus* sp A, *Daemonorops* sp B and *Korthalsia bejaudii*. *Calamus oligostachys* is only known from two sites close together.

Rattan species richness is much lower in Indochina than in Malaysia and Indonesia. For example, Peninsular Malaysia covers an area 55% that of Lao but it has over 100 rattan species in nine genera.

Habitat

We know very little about the way Lao rattans live their lives because so little research has been done. Our observations during collecting trips allow us to make some general comments about habitat preferences.

Most Lao species are found in various types of evergreen forest. Climbing species are often most numerous in old tree fall sites or along streams where there is extra light available. Some occur mainly in unlogged forest whilst others seem to do well in degraded forest or even tall evergreen scrub. Some species are absent from logged areas, either because they can't grow there or because they were over-harvested during the period of logging.

A few species seem to be associated with other habitats. *Calamus acanthophyllus* grows only in deciduous forest and dry scrub which has regular fires. It has special adaptations to survive fire. *C. viminalis* does occur in evergreen forest but can also be found in the moister parts of deciduous forest and is most typical of dry scrubby habitats near villages. *C. tenuis* and *C. godefroyi* seem to be species of lowland floodplains and although these have mostly been deforested the two species still grow around Vientiane in patches of scrub, especially if there is frequent flooding. A few species have been recorded in forest on karst limestone mountains, including *C. henryanus*, *C. solitarius*, *C. laoensis* and *C. flagellum*.

Each Lao species has a preferred range of altitudes. Most appear to prefer lower altitudes and the richness of species becomes less and less above about 600 m. Beyond 1300 m few species have so far been recorded, although fieldwork at these altitudes has been limited. Species typical of the highest altitudes probably include *C. acanthospathus*, *C. nambariensis* and *Plectocomia himalayana*. All altitudes in this book are approximate, to the nearest 50 or 100 m.

Flowering and fruiting

We know the flowering and fruiting seasons quite well for a few species which grow close to Vientiane but for most species we have only a few specimens and some vague reports to pinpoint the times of flower opening and fruit-ripening. This is an aspect for which more published information would be very valuable, since knowing when ripe fruit can be found is crucial to planters.

The information is presented in an unsummarised format, giving the months (as numbers from 1, January to 12, December) when flowers (in bud or open) and fruit (unripe or ripe) have been found in the various countries of Indochina. We think this is less misleading than guessing the seasons or giving vague summaries of the data. Readers can see what is known, add their own observations and draw their own conclusions. Information from outside Indochina was not included because the climate pattern differs in those areas. Readers should be aware that there may not be uniform flowering and fruiting seasons within Indochina either, particularly between Lao and Vietnam, due to the different times of the rainy seasons.

The limited evidence shows the following three features. Firstly, most rattan fruits take many months to ripen (for example, around Vientiane *Calamus viminalis* flowers in August-September and ripens in March-April of the following year). Secondly, some species may take more than twelve months for the fruit to ripen (e.g. *C. erectus* and *C. rudentum*). And thirdly, although many individual species seem to have definite flowering and fruiting seasons, these differ widely between species. For example, *C. siamensis*, although closely related to *C. viminalis*, seems to flower in March or April and ripen some time after July.

Uses

In the species accounts a simple classification has been made of the suitability for use under four headings: handicrafts, trade, food and other. This was mainly based on the opinions of experienced rural people during specimen collection. It would be worthwhile to gather the opinions of traders too, and to try to relate their trade names to botanical names.

The stems of climbing species can often be used for handicrafts or making furniture. There was a large export market in the past but exports of raw canes are now discouraged (with some exceptions). Handicrafts are exported to Thailand, France and probably other countries. Stem appearance and quality differ for each species. Flexibility, attractive colour and evenness of diameter are among the important characteristics sought by traders but rural people are less selective for many household uses. *Calamus poilanei* is outstanding among the larger diameter canes and is the preferred species for making furniture frames. Preferred small-diameter species in central Lao include *C. gracilis* and *C. solitarius*. The species sought by traders are usually the best of the abundant species – it may be that even higher quality stems exist amongst the rarer species and these should also be considered as candidates for cultivation.

Rattan shoots are eaten throughout the country but, apart from northern and eastern Thailand, locally in Vietnam and in parts of Borneo, are rarely eaten elsewhere in Asia. Most species taste bitter, although *C. rhabdocladus* is quite sweet. The nutritional value of rattan shoots has apparently not been studied. Shoots, probably mostly of *Daemonorops jenkinsiana*, appear in very great numbers in the markets of northern Lao and many other species are sold in moderate quantities elsewhere in the country. They are almost all gathered in the wild, but cultivation is spreading (see page 89). There is a small-scale export market serving Lao and Thai people in Europe and America and, at least occasionally, consumers in Yunnan. Exported shoots are mostly sun-dried in Lao, but in Thailand many are canned.

Other minor uses include the sale of edible fruits in local markets, thatching of houses (*Daemonorops jenkinsiana* is widely used for this in northern Lao) and the sale of decorative seeds, again of *Daemonorops jenkinsiana*, to Vietnam and China for handicrafts. We have only found two recorded medicinal uses for Lao rattans (*C. acanthophyllus* roots have been used as a malaria cure and a treatment in childbirth) but there may be others we do not know about.

Field keys

Before using these keys familiarise yourself with the main parts of a rattan described on pages 81-87. Then look carefully at the rattan you want to identify. They are not simple plants. They are long-lived and the climbing species have a complicated life history because their rapidly extending stems start life in the moist, shady forest understorey then climb into hot, brightly sunlit areas high in the canopy. If a tree or branch falls the rattan may drop back into the shade and have to begin climbing again. Big, old clumps have often been affected by harvesting, disease or other catastrophes. Some species (in the genera *Korthalsia*, *Myrialepis*, *Plectocomiopsis* and *Plectocomia*) also experience a great change when the stem stops extending and prepares for a single massive flowering effort before dying.

All this means that different parts of the same plant can look very different. You need to look at all parts of the plant and apparently similar plants nearby to gain an understanding of which features are unvarying and which have a wide range of variation, depending on age, environment and genetic differences. Concentrate on the upper 'adult' part of the stem rather than seedlings and young shoots.

There is an initial checklist of features which you should note for any rattan. The first four points are given special prominence at the top of each species account:

- **is it stemless, erect or a climber?**
  A young climber may look erect, but should have long internodes with tubular sheaths. Climbing organs should appear by the time the stem is 2-3 m tall.
- **is it clustering or solitary?**
  Check that all the stems come from one plant and not two close together. The first stem may be a few metres long before it produces new shoots.
- **what climbing organ does it have, if any?**
- **is there a knee and is it prominent?**

The next thing to look at is the **arrangement, shape, size and colour of the leaflets**. With all this information you should already be quite close to a successful identification.

First look at the list of rare special features on page 11. These may help you get close to the correct name in one step. The next stage is to work through the keys, checking additional features of the plant as required. This should lead you to one or two probable names. From there go to the relevant species accounts and decide whether the more detailed information all fits the plant you are looking at. Check the list of similar species under **'Compare'**.

If the description sounds wrong you may be looking at another species, so read the accounts for all the similar species or try going back to the keys. If there are points where you are not sure which option to choose, try all of them. If you still don't feel confident that you know what the species is, leave it unidentified. Don't guess! It could be an unusual individual or a species not in the book.

When you begin, identification will take a long time but as you get to know each species you may be able to identify it without the keys, simply by remembering the 'character' of each species – the distinctive combination of colour, shape, texture and size. These things are as difficult to describe in words as your best friend's face; the only way to learn them is to see them. Don't forget to collect herbarium specimens to add to our knowledge. See page 87 for details of how to make a specimen.

Special features of some rattan species

Numbers refer to species accounts. Brackets mean the feature is only present sometimes.

Habit and climbing organs
No stem: 1
Stem erect, not climbing: 2-2b, 3-3b, 4a
Climber without climbing organ: 17, 17a
Flagellum only: 4, 5-15a (not 8a), 18-21
Cirrus only: 8a, 22-31a
Short cirrus and flagellum: 16
Stem solitary: 2, 8, (9), 12, possibly others

Inflorescence
Only 1-branched: (1), (2), (2a), (13)

Ocrea
Long, papery: 1, 4a, 7, 29
Bristly collar (≤2 cm): 5a, 9, 15, 15a, 17a, 18
Lacerate, spiny collar or a pair of 'ears' (2-5 cm): 2, 2b, 3, 4a, 19
Net-like: 27-27b

Upper leaves
Petiole very short (≤3 cm): (5), (6a), 9, 10, (12), (13), (15a), (29), 30-30b, 31
White or grey indumentum under leaflets: 2a, 6a, 17, 17a, 18, 27-27b, 30-30b, 31a
Leaflet veins stop at the edges, not the tip: 9b, 27-27b
Leaflets diamond shaped: 27-27b
Leaflets end in a hair-like tip (≥5 cm): 28, 29, 31

Habitat
Planted for shoot production in Lao: (4), (5), 6, (26)
Grows in dry, open habitats: 1, 4, 26
Grows in very wet places: 6, 6a
Can grow above 1800 m: 2b, 9, 23, 31

How the keys work:

Each column gives you a choice of two or more alternatives. Start in the first column of Field Key A and choose the row with the alternative that fits your rattan best. This eliminates all the other options, and so all the other rows of the table. Go across to the right. This may take you to another choice. At each choice take the row which fits best and eliminate the others. When you reach the last column you will find the names of one or two species, or perhaps you will be taken to another, more detailed key.

Some species appear in more than one place because they have more than one arrangement of leaflets. The keys are designed for people working in the field so we have not used the appearance of the inflorescences unless it is necessary – they are sometimes hard to find or too high in the trees to reach.

Example

1. We have found a rattan in the forest. We go to the first column of Field Key A. It has no climbing organ, so we go to the right from here, eliminating the rest of the table.
2. In column 2 we see that the option 'stem erect, stiff, ≥0.5 m tall' fits best. This doesn't identify the species but instead leads us to a new key, Field Key B.
3. In column 1 of Key B we see that 'leaflets not white below' fits our plant best, so we choose the second option.
4. In column 2 we choose the first option, because the leaflets are about 70 cm long.
5. Our last choice is easy because the petiole clearly has complete rings of 10 cm long spines around it – so this is probably *Calamus erectus*, species 3.
6. To check whether this is right we look at the species account. The photograph, measurements and other details make it clear that we have picked the right species.

Field Key A – General

no climbing organ on sheath or leaf	no stem		*C. acanthophyllus* 1
	stem erect, stiff, ≥0.5 m tall		see Field Key B
	stem climbing, thin and flexible	ocrea smooth, leaflets smooth on the faces, last pair of leaflets ≥5 cm	*C. hypoleucus* 17
		ocrea bristly, leaflets spiny, last pair ≤4 cm	*C. salicifolius* 17a
cirrus and flagellum, both ≤20 cm			*C. bimaniferus* 16
only a flagellum, ≥30 cm	leaflets strongly grouped or irregular		see Field Key C
	leaflets regularly pinnate		see Field Key D
	leaflets interrupted pinnate		see Field Key E
only a cirrus, ≥30 cm	knee absent	leaflets lanceolate	see Field Key F
		leaflets diamond-shaped	*Korthalsia* spp 27-27b
	knee present		see Field Key G

Field Key B – Erect species

leaflets white below				*C. oxycarpus* 2a
leaflets not white below	leaflets ≥50 cm	petiole spines ≥5 cm, in neat rings all around petiole		*C. erectus* 3
		petiole spines ≤3 cm, solitary or in pairs		*C. dongnaiensis* 3a
	leaflets ≤40 cm	leaflets regular, all lying in one plane		*C. harmandii* 2
		leaflets grouped, all lying in one plane	leaflets up to 24×1 cm, rachillae curved back, fruit with long conical point	*C.* sp A 2b
			leaflets up to 40×2 cm, rachillae pressed against axis, fruit spherical	*C.* sp B 3b
		Leaflets grouped, pointing in many different planes		*C. thysanolepis* 4a

Field Key C – Climbers with flagellum only, leaflets strongly grouped or irregular

Leaflets plumose, lying in many different planes	Sheath ∅ ≥3.5 cm, densely covered in flat spines			*C. laoensis* 18 and *C. longisetus* 18a
	Sheath ∅ ≤2.5 cm, with scattered slender spines			*C. viminalis* 4
Leaflets not plumose, lying in one plane	Leaflets irregularly arranged	Petiole ≤3 cm, leaflets ≤20 cm long, sheath spines with crescent-like bases		*C. godefroyi* 6a
		Petiole ≤15 cm, largest leaflets ≥40 cm, sheath spines needle-like.		*C. kingianus* 15
	Leaflets neatly grouped, usually in 2s on each side, with a terminal group of 4	Ocrea bristly, upper surface of leaflet with three lines of bristles, last 2 leaflets not joined.		*C. dioicus* 13
		Ocrea without bristles, upper surface of leaflet with few bristles, last 2 leaflets joined 20-60%	Clustering First partial inflorescence ≤11 cm long, barely or not branched	*C. oligostachys* 12
			Solitary stem First partial inflorescence ≥40 cm long, well branched	*C. solitarius* 13
			Clustering First partial inflorescence 10-40 cm long, well branched	*C. tetradactylus* 14

Field Key D – Climbers with flagellum only, leaflets regularly pinnate

<table>
<tr><td rowspan="2">Leaflets oval or diamond-shaped
Stem solitary or clustering</td><td colspan="3">Petiole 1-3 cm, leaflet veins finish at the leaflet tip, ocrea spiny.</td><td>C. acanthospathus 9</td></tr>
<tr><td colspan="3">Petiole 6-20 cm, veins finish along leaflet margins, ocrea not spiny</td><td>C. bousigonii 9a</td></tr>
<tr><td colspan="4">Leaflets linear (much longer than wide)
Stem solitary</td><td>C. poilanei 8</td></tr>
<tr><td rowspan="7">Leaflets linear or lanceolate (much longer than wide)
Stem clustering</td><td rowspan="2">Sheath ∅ ≥3 cm</td><td colspan="2">Sheath spines black, leaflets with three lines of bristles above, petiole spiny above</td><td>C. rhabdocladus 21</td></tr>
<tr><td colspan="2">Sheath spines paler at the base, leaflets with one line of bristles above, petiole smooth above</td><td>C. flagellum 19 and
C. rudentum 20</td></tr>
<tr><td rowspan="5">Sheath ∅ ≤2.5 cm</td><td colspan="2">Ocrea forms papery tassels up to 10 cm</td><td>C. guruba 7</td></tr>
<tr><td colspan="2">Ocrea a short bristly collar</td><td>C. walkeri 5a</td></tr>
<tr><td rowspan="3">Ocrea tiny, unarmed</td><td>Spines yellow with red tips, petiole ≥7 cm</td><td>C. siamensis 5</td></tr>
<tr><td>Spines green with black tips, petiole ≥8 cm</td><td>C. tenuis 6</td></tr>
<tr><td>Spines green with black tips, petiole ≤3 cm</td><td>C. godefroyi 6a</td></tr>
</table>

Field Key E – Climbers with flagellum only, leaflets interrupted pinnate

Sheath ∅ ≥4 cm, thickly covered in long, dark, paper-thin, triangular spines			*C. rhabdocladus* 21
Sheath ∅ ≤3 cm, with other types of spines	Petiole 0-3 cm. Spines ≤0.5 cm with swollen bases and short points		*C. gracilis* 10
	Petiole 5-25 cm Spines ≥1.5 cm, without swollen bases and with long points	Some sheath spines broad, brownish green with lots of brown indumentum. Inflorescence branches pressed against the axis	*C. henryanus* 11
		Sheath spines slender, yellow-brown with red-tips, bases curved, little indumentum. Inflorescence branches spreading from the axis	*C. siamensis* 5

Field Key F – Climbing species with cirrus, knee absent

Ocrea ≥5 cm long, papery Leaf surface with golden bristles				*Plectocomiopsis geminiflora* 29a
Ocrea ≤0.5 cm long Leaf surface without bristles	Lower sheaths with broad spines ≤5cm long, in combs			*Myrialepis paradoxa* 28
	Lower sheaths with needle-like spines ≤3 cm, scattered or in combs	Leaflets never white below, with a hair-like tip ≥5 cm.		*Plectocomia himalayana* 31
		Leaflets initially white below, without a long hair-like tip	Petiole ≤3 cm, sheath spines scattered or in rows of ≤10. Fruit not bristly	*Plectocomia pierreana* and close relatives 30-30b
			Petiole ≥10 cm, sheath spines in neat rows of ≥ 10. Fruit bristly	*Plectocomia elongata* 31a

Field Key G – Climbing species with cirrus, knee present

Sheath spines slender or triangular and thick, green, red or yellow Inflorescence erect with widely spread branches and narrow tubular bracts which never fall off **NOTE – species 22-25 are variable and can be difficult to identify. Read the species accounts carefully.**	All leaflets strongly grouped Sheath spines mixed long/short, ≤3 cm Fruit ≤1.2 cm long, perianth not remaining tubular, scales unchannelled	*C. palustris* 22
	Leaflets strongly grouped in the middle of the rachis, sometimes regular at the base Sheaths with many short and long spines up to 4 cm Fruit ≥2.5 cm long, perianth remains partly tubular, scales channelled	*C. nambariensis* 23 and *C. obovoideus* 23a
	All leaflets strongly grouped Sheath spines all or nearly all long, up to 9 cm Fruit ≤2.2 cm long, perianth remains partly tubular, scales channelled	*C. platyacanthus* 24
	Leaflets perfectly regular Some lower sheaths with stripes of white indumentum and all spines tiny and swollen like the teeth of a rasp. Longest spines ≤4 cm. Fruit size ≤2.0 cm long, perianth remains partly tubular, scales channelled.	*C. wailong* 25
Sheath spines triangular, thick,bent down. Inflorescence long with a pendulous, clawed axis.		*C. ceratophorus* 8a
Sheath spines flat, paper-thin, black Inflorescence short with crowded branches, covered by big overlapping boat-shaped bracts until the flowers open		*Daemonorops jenkinsiana* 26
Sheath spines flat, paper-thin, straw coloured. Inflorescence longer, each partial inflorescence covered by its own bract which falls off before flowering		*Daemonorops* sp A and sp B 26a & b

The species accounts

There are 31 main species which have detailed double-page accounts and 20 similar 'allied species' which have short accounts in the 'Compare box', bottom right. Allied species are mostly those which occur outside Lao, or those for which we have little information. They resemble the main species in most aspects except those specifically noted as different. The 'Compare' box also suggests other species that might be confused with the main species, and the key features for separating them.

Below the scientific name of the main species we list all the local names which we have heard used for this species in Lao. There are probably many others which we have not recorded. See page 6 for more notes on local names.

The next line lists the four most important features you should note when you look at an unidentified rattan (see page 10 and 'Parts of a Rattan' pages 81–87).

The 'Look for' box tells you the combination of features which is special to this plant.

Next we give information on habitat (see page 8), fruiting and flowering (page 9), uses (page 9) and planting (page 89), all specifically for Indochina.

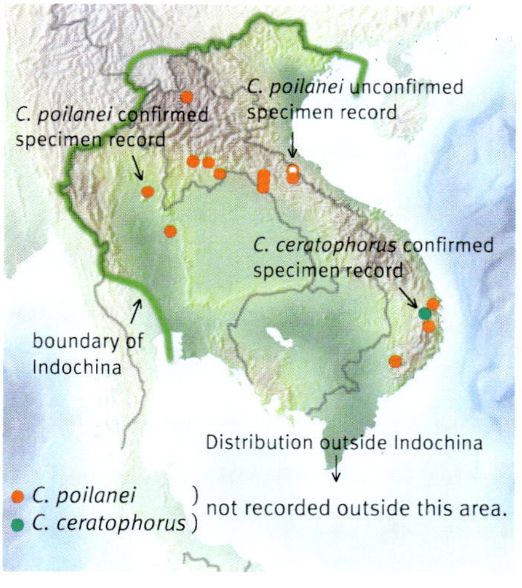

On the map each species has dots of a different colour. A closed symbol marks a locality where a herbarium specimen of this species has been collected. We use a few open symbols for important records where we have not confirmed the identification. This may be because we did not see the specimen or because the specimen we saw was incomplete.

The dark green line encloses Indochina, the area where we map all records. Outside Indochina we don't map the records, we just list the main areas the species has been found (see page 7).

The photo includes the code number of a specimen taken from the same plant and kept at Kew. This makes the picture more useful in the future if the names of the species change. The line drawings are often based on more than one specimen, so we do not give a code number. The drawings of the rachillae and fruits are all life size although some rachillae are longer than the box so only a part is shown. The leaf and inflorescence drawings are at varying scales; an outline of a 20 cm long hand or a normal sized man is included in each picture to help you visualise the true size.

Measurements and descriptions of the parts of the plant are almost all based on specimens and field observations from Indochina. Remember that plants that are young, sick or in deep shade can sometimes have parts smaller than the ranges given here, but parts which are larger than the maximum listed are much rarer.

1 *Calamus acanthophyllus*

(wai tia, wai kok, wai foom, wai nang)

No stem Clustering No climbing organ No knee

Look for:

- Stemless habit and open, not tubular leaf sheaths.
- Leaves erect, the leaflets often in groups and with stiff golden spines.

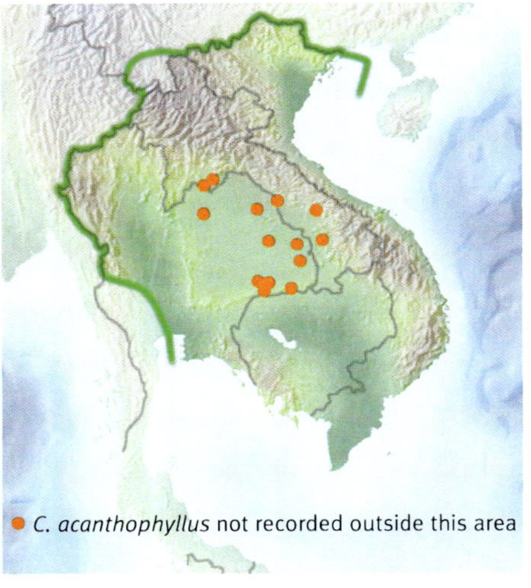

Habitat
In Lao grows in dry, open forest and bamboo scrub, often in areas with regular fires, below 200 m. Habitat similar in Thailand.

Fruiting and flowering months
Flowers in bud 7, 11 (Thailand), open 3 (Lao). Fruit unripe 1, 4 (Lao), 12, 6 (Thailand), ripe 11 (Thailand).

Uses
Handicrafts: leaves used to make brooms.
Trade: not suitable.
Shoot: edible.
Root can be used to treat malaria.

Planting
None known in Lao.

Stem
Spines yellow, conical at base, 0.1–0.5 cm long.

Ocrea a pair of papery, red-brown ear-like projections.

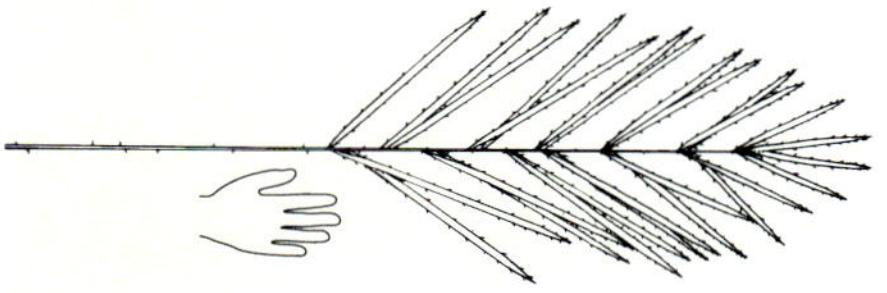

Whole **leaf** 0.5–1.5 m long.

Petiole of upper leaves 10–90 cm long.

Largest leaflets 15–23 cm × 0.8–1.5 cm.

Leaflets plumose, usually in groups of 2–3. Above three prominent veins, the middle one with stiff, erect yellow spines. Similar erect spines on margins.

Inflorescence 0.4–1.1 m long with no flagellum. Erect.

Primary bracts entire, slightly inflated.

♂ usually 1–2-branched.

♀ usually 1-branched.

Fruit

1.2 cm long

0.9 cm wide

Scales dull whitish (yellow when dry) with fine dark margins.

Seed not ruminate.

Compare:

- *C. salicifolius* (17a): short climbing stem, tubular leaf sheaths, leaflet spines black if present.
- *C. viminalis* (4): seedling similar, but has tubular leaf sheaths and long, straight spines on the rachis.

2 *Calamus harmandii*

(nya seui, yeu mo)

Erect Solitary No climbing organ No knee

South Lao is the only place in the world where this species is known.

Look for:

- Erect stem 0.5–1.1 m tall with open, not tubular sheaths.
- Leaflets narrow, regularly arranged, ≤40 cm long.
- Inflorescence slender, erect, very little branched, almost unarmed.

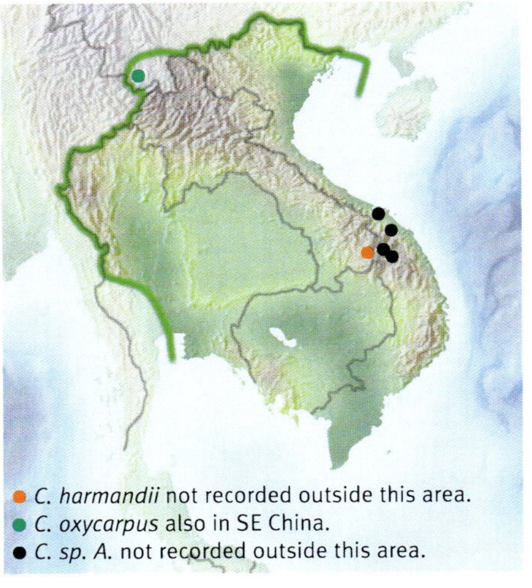

Habitat

In Lao grows in evergreen forest at 150–200 m. Only known from one site in the world.

Fruiting and flowering months

Flowers open 5 (most already fallen).
Fruit unripe 3, 5.

Uses

Handicrafts: not suitable.
Trade: not suitable.
Shoot: too small to eat.

Planting

None known in Lao.

KP 398/400

Stem

Stem up to 1.1 m long and 4 cm diameter.

Spines at base of petiole straight, up to 3 cm long.

Ocrea forms two tattered bristly 'ears'.

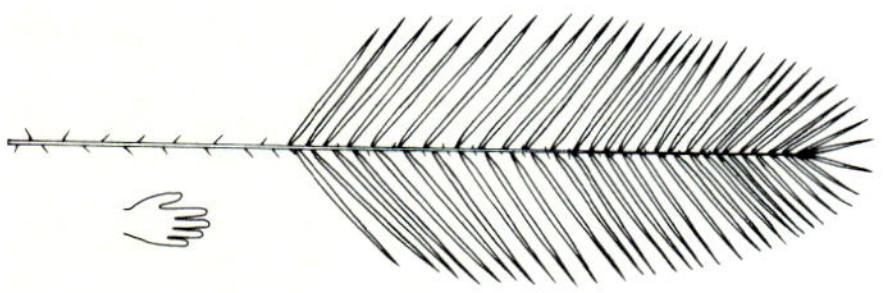

Whole **leaf** 1.4–1.7 m long.

Petiole of upper leaves 35–85 cm long.

Largest leaflets 40 cm × 1.7 cm.

Leaflets with three prominent, bristly veins on the upper surfaces, below naked. Margins bristly.

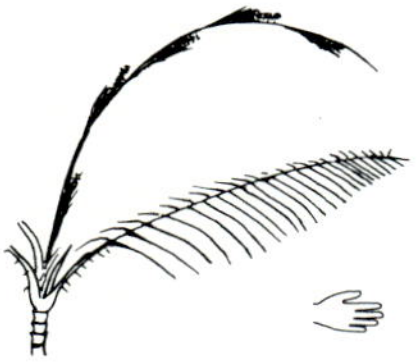

Inflorescence 1.3–1.4 m long with no flagellum. Erect. Branches pressed against the axis.

Primary bracts lacerate, not inflated, almost unarmed.

Male and female flowers all very crowded, not in neat rows.

♂ usually 2–branched but the 2nd-order branches hard to see because they are all pressed together.

♀ usually 2-branched but looks 1-branched because the 2nd order branches are stuck together.

Fruit

Mature fruit and seed unknown.

Compare:

- *C. acanthophyllus* (1): clustering, stemless, with grouped leaflets.
- *C. erectus* (3): clustering, 5–6 m tall, spines in combs around petiole, leaflets much larger.
- *C. dongnaiensis* (3a): leaflets broad, can be grouped, inflorescence well-branched, many spines, rachillae pendulous.

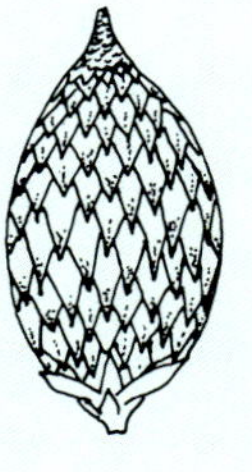

• 2a *Calamus oxycarpus*
Erect, 2–3 m tall. Leaflets grouped, 25–30 × 1.5–4 cm, with white indumentum below. Inflorescence only 1-branched, flowers in neat rows. Fruit large, (up to 30 × 13 mm), dark brown and orange when dry, with a long conical beak and 24 rows of scales (see picture).

• 2b *Calamus* sp A
Leaflets strongly grouped, up to 24 × 1 cm. Sheath and petiole spines flat, black, in neat rows (like a tiny *C. rhabdocladus* **21**). Inflorescence up to 80 cm, well branched, primary bracts lacerate, rachillae curved back, the flowers in neat rows. Flowers in bud 11. Grows in evergreen forest at 1200–2000 m.

3 *Calamus erectus*

(wai namsay, blong chintra, kok tamarr)

Erect Clustering No climbing organ No knee

Look for:

- Erect habit and very deeply split leaf sheath with a pair of bristly 'ears'.
- Petiole spines long, golden, joined in combs running around the petiole.
- Fruit very large and dark on erect, well-branched inflorescences.

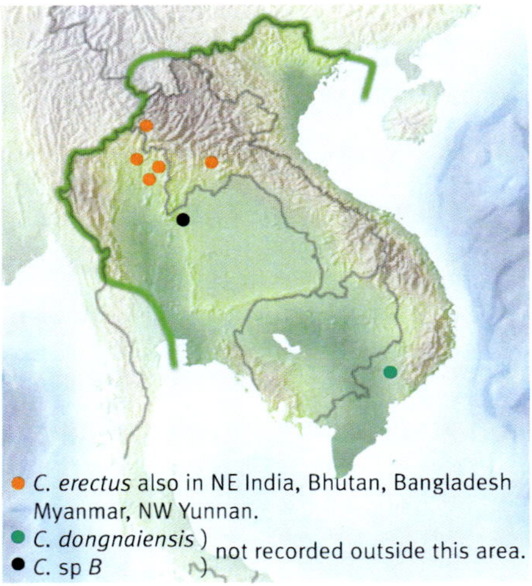

Habitat

In Lao grows in evergreen forest at 450 m and probably also higher. In Thailand recorded at 600–1600 m.

Fruiting and flowering months

Flowers open 2 (Thailand).
Fruit ripe 3, 4 (Lao), 1–2 (Thailand). May take ≥1 year to ripen.

Uses

Handicrafts: not suitable.
Trade: not suitable.
Shoot: edible.

Planting

None known in Lao.

TDE 69

Stem

Stem up to 6 m tall and 3–4 cm diameter.

Sheath 5–7 cm diameter.

Sheath hardly tubular, dull green-brown.

Spines many, flat, 2–5 cm long, in lines, with small black bristles to either side.

Ocrea over 15 cm, covered in short bristles. Begins as a tube, soon splits to form an 'ear' on each side of the petiole.

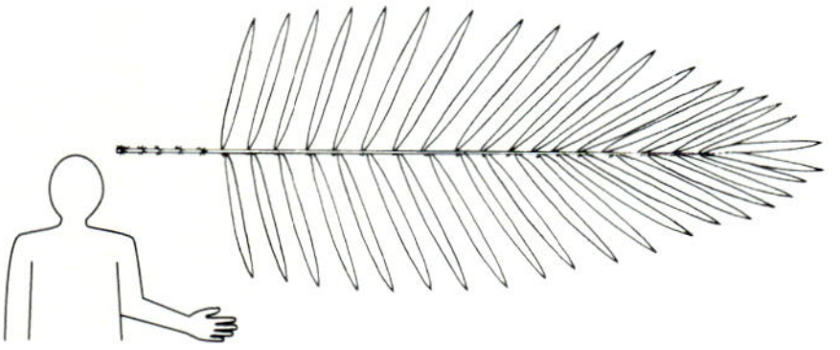

Whole **leaf** 3–5 m long.

Petiole of upper leaves 50–150 cm long. Petiole spines 5–15 cm, joined in combs running around the petiole.

Largest leaflets 60–80 cm × 3.5–5 cm.

Above one (rarely 2–5) veins prominent, with long or short bristles. Below, central vein may also have bristles. Margins bristly. Many large younger leaves are irregularly pinnate.

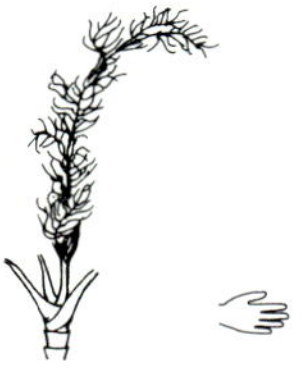

Inflorescence 0.5–1 m long with no flagellum. Erect, amongst petioles.

Primary bracts highly lacerated, covering base of partial infloresences.

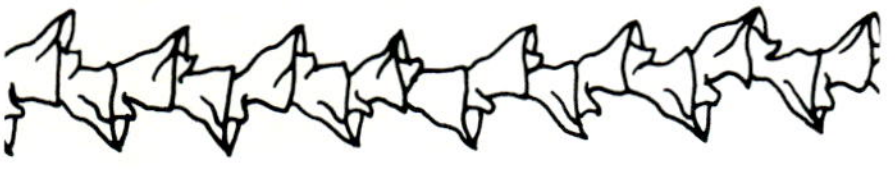

♂ usually 2-branched.

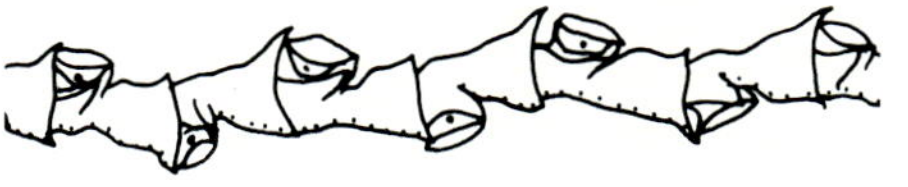

♀ usually 2-branched.

Fruit

3–3.5 cm long

2–2.5 cm wide

Reddish brown with dark and light patches, deeply channelled.

Seed ruminate.

Compare:

- *C. laoensis* (18), *C. flagellum* (19), *C. rudentum* (20) and *C. rhabdocladus* (21): all climbing.
- *C. thysanolepis* (4a): leaflets in groups of 2–6 on each side of the rachis, lying in many planes, ≤45 cm long. Rachis spines solitary. Rachillae much thinner, fruit ≤1 cm diameter

● 3a *Calamus dongnaiensis*
Erect, 1–3 m tall, clustering. Leaflets up to 60 × 3 cm, can be grouped. Petiole and rachis spines solitary. Inflorescence long, slender, drooping, with slender rachillae.

● 3b *Calamus* sp B
Leaflets strongly grouped, up to 40 × 2 cm. Rachis spines flat, black, in groups. Female rachillae 1–3 cm long, very thick, pressed against the branches. Fruit globose, 1.5 cm diameter, scales dark, unchannelled. Seed ruminate. Fruit unripe 2, 5 (Thailand). In evergreen forest at 1300–1550 m.

4 *Calamus viminalis*

(wai ton, wai nyair, wai na, wai khom, wai namhang, wai keethao, wai nang, wai tiudeet, wai namleuang, rebou, blong chang, katengparua)

Climbing Clustering Flagellum Knee conspicuous

A very common species, often surviving as old clumps around villages.

Look for:

- Leaflets in groups of 2–4, pointing in many different planes, with one line of pale spines above.
- Sheaths spines long, yellow with red tips.
- Petiole has long, straight spines (up to 5 cm long).

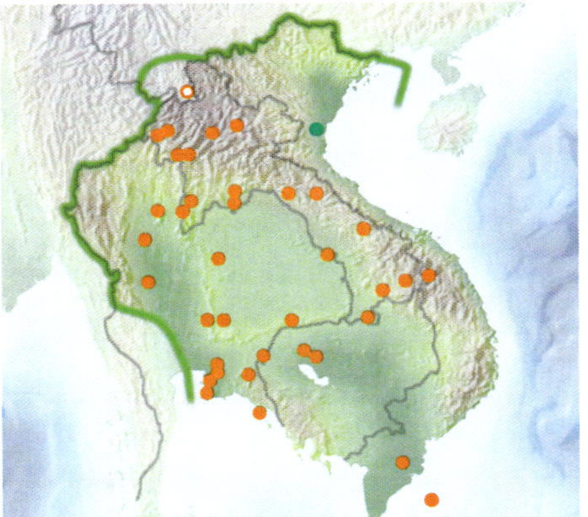

● *C. viminalis* also in NC, NE, S India, Andamans, Bangladesh, Myanmar, S and SW Thailand, Peninsular Malaysia, Java, Bali and ?NW Yunnan.
● *C. thysanolepis* also in SE China.

OL 220-222

Habitat

In Lao and Thailand in scrub/forest edge at 100–600 m. In N also evergreen forest up to 600 m.

Fruiting and flowering months

Flowers in bud 8 (Lao), 9–12 (Thailand), open 8–10 (Lao), 7, 9, 12 (Thailand). Fruit unripe 9–3 (throughout), ripe mainly 4, sometimes 3–6 (Lao), also some very unripe 5–6 (Thailand). Two flowering seasons?

Uses

Handicrafts: moderate quality.
Trade: suitable.
Shoot: edible.
Fruit sometimes sold for food.

Planting

Plantation at Nam Xuang. Managed gardens for shoots (Bokeo Prov.) and canes, shoots and fruit (Vientiane Prov.) Widely planted for shoots in NE and E Thailand

Stem

Stem up to 35 m long and 0.5–1.5 cm diameter, occasionally even less. Sheath 0.9–3 cm diameter (rarely even 0.5 cm).

Sheath pale green with white indumentum which rapidly falls off.

Spines yellow with red tips when alive, needle-like with a broader base, up to 4 cm long.

Ocrea tiny, dry, with no spines or bristles.

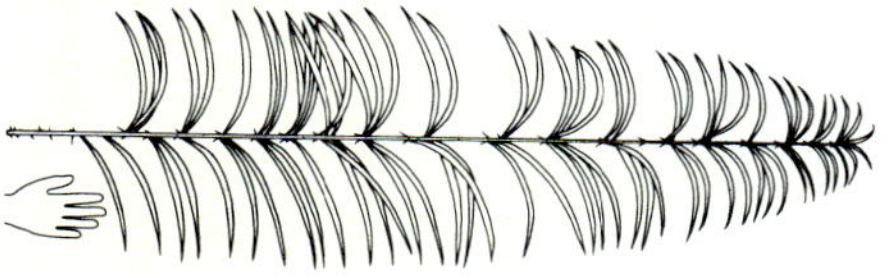

Whole **leaf** 1.0–1.5 m long.

Petiole of upper leaves 5–20 cm long with straight spines 1–4 cm long.

Largest leaflets 15–35 (rarely only 7–10) cm × 1–1.5 cm.

Leaflets in groups of 2–4 on each side, pointing in all directions. Above middle veins prominent. One line of yellow-based bristles above and below. Margins bristly.

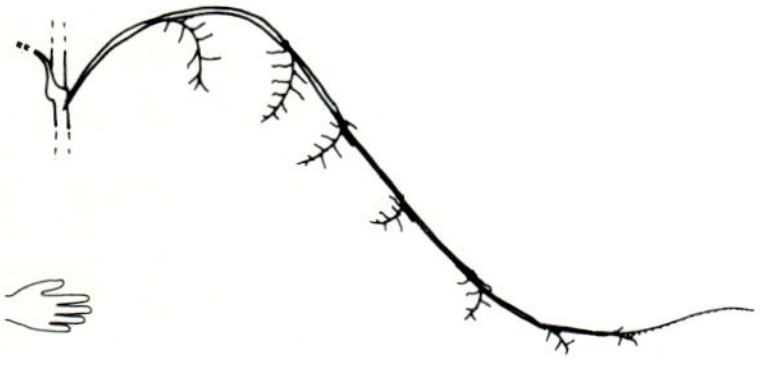

Inflorescence 1.5–3 m long plus flagellum. Branches quite stiff, slightly pendulous.

Primary bracts entire, closely sheathing, with scattered claws.

♂ usually 3-branched but looks 2-branched. Rachillae very short (≤1 cm), crowded with tiny flowers. Picture shows a branch with many rachillae.

♀ usually 2-branched. Rachillae 5–15 cm long. Some rachilla bracts contain two female flowers with one neuter flower, others only one. The percentage varies between plants.

Fruit

0.8 cm long

0.8 cm wide

Whitish (yellow when dry) with fine dark margin. Perianth wholly split, no tubular part.

Seed not ruminate.

Compare:

- *C. acanthophyllus* (1): Leaves similar but smaller, leaflets spinier. No stem.
- *C. salicifolius* (17a): Smaller, no climbing organ, small bristly ocrea, inflorescence much smaller.
- *C. siamensis* (5): Very similar but leaflets in one plane, interrupted pinnate or regular.

● 4a *Calamus thysanolepis*

Leaflets grouped and plumose, but stem erect, no knee or climbing organs. Leaf bases with needle-like black spines and a long papery ocrea. Inflorescence quite short, primary bracts strongly lacerated. Grows up to 300 m altitude. Fruit ripe 10 (Vietnam).

5 *Calamus siamensis*

(wai khom, wai nam, wai deng, yo, re dark)

Climbing Clustering Flagellum Knee conspicuous

Widely cultivated in Thailand but not yet planted in Lao.

Look for:

- Sheath spines yellow with red tips when alive, ocrea tiny and not bristly.
- Leaflets very closely set in one plane, regular or interupted pinnate in groups of 2–10.
- Leaflets with one line of pale based spines above.

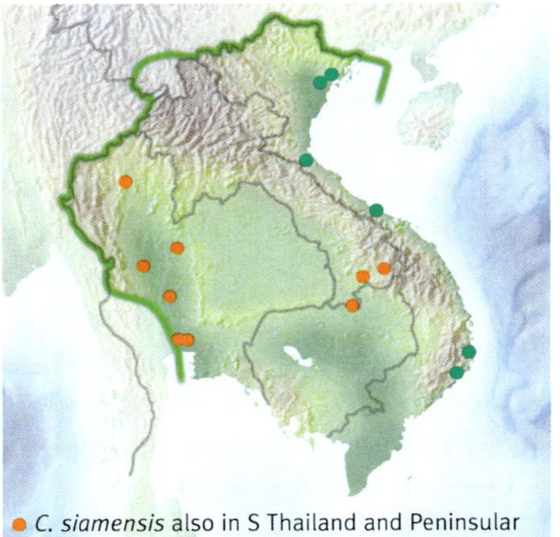

● *C. siamensis* also in S Thailand and Peninsular Malaysia.
● *C. walkeri* also in SE China.

Habitat

In Lao grows in scrub forest and degraded semi-evergreen forest at 100 m. In Thailand recorded at 0–300 m.

Fruiting and flowering months

Flowers in bud 3-4 (Thailand).
Fruit unripe 5 (Lao), 3-6 (Thailand).

Uses

Handicrafts: suitable.
Trade: suitable.
Shoot: edible.

Planting

None known in Lao. High potential. Widely planted for shoot production in NE Thailand.

KP 401

Stem

Stem up to 20 m long and 0.9–2 cm diameter.

Sheath 1.2–3 cm diameter.

Sheath greenish, sometimes with grey-brown indumentum.

Spines brownish to yellowish brown with reddish tip, 1–2.5 cm long, often with crescent bases.

Ocrea tiny, dry, with no spines or bristles.

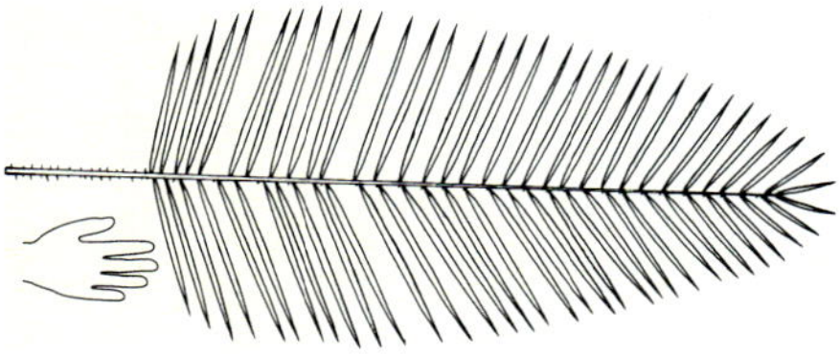

Whole **leaf** 0.8–1.2 m long.

Petiole of upper leaves 7–22 cm long, sometimes ≤5 cm on sheaths with inflorescences.

Largest leaflets 11–33 cm × 1–2 cm. Leaflets regular on some leaves, interrupted pinnate on others. Above 1–3 prominent veins with pale-based bristles on one. Below sometimes a few bristles. Margins bristly. Rachis has straight spines 1–3 cm long below.

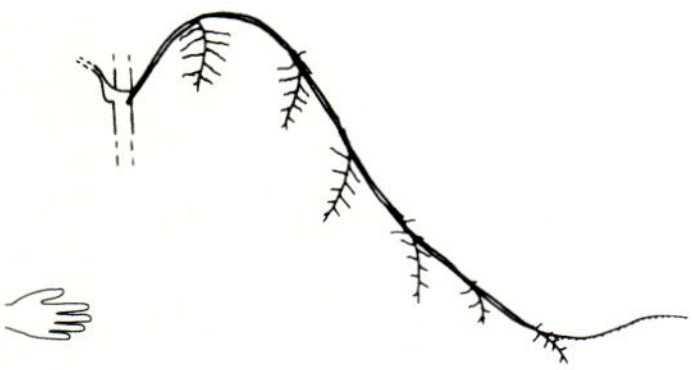

Inflorescence 0.6–1.6 m long plus flagellum. Branches pendulous.

Primary bracts entire, closely sheathing.

♂ usually 3-branched but looks 2-branched. Rachillae ≤1 cm, crowded with tiny flowers. Picture shows a branch with many rachillae.

♀ usually 2-branched. Usually 2 female flowers with each neuter flower.

Fruit

0.8 cm long

0.8 cm wide

Scales whitish (yellow when dry). Perianth wholly split, no tubular part.

Seed not ruminate.

Compare:

- *C. viminalis* (4): leaflets in groups of 2–4 on each side, pointing in many different directions. Often in dry, open habitats.
- *C. tenuis* (6): leaflets regular with several lines of black hairs above, sheath spines green with black tips, male rachillae longer, fruit perianth remains partly tubular.
- *C. godefroyi* (6a): Leaflets regular, grey-white beneath, petiole ≤2 cm, top side of rachis has small spines, sheath spines green with black tips, fruit perianth remains partly tubular.

• 5a *Calamus walkeri*

Ocrea 0.5–1.5 cm, very bristly. Leaflets with three lines of dark bristles above, and can reach 50 cm long. Dry specimens often a cold, bluish-green colour. Sheath spine bases swollen, not crescent-shaped. Male rachillae tiny. Fruit perianth with slight tubular part. In forest up to 300 m. Flowers in bud 1, 5 (Vietnam).

6 *Calamus tenuis*

(wai nyair, nyair)

Climbing Clustering Flagellum Knee conspicuous

The main species planted for commercial shoot production in Lao.

Look for:

- Leaf sheath with green, black-tipped spines which have downswept, crescent bases.
- Leaflets regularly arranged, with several lines of black bristles above.
- Petiole ≥8 cm, ocrea almost absent.

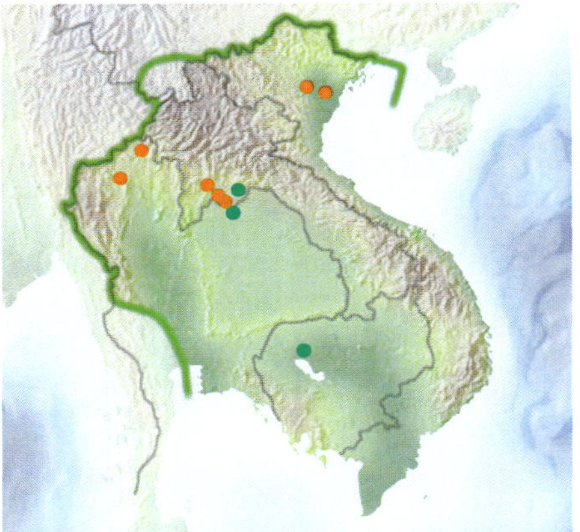

● *C. tenuis* also in NC and NE India, Bangladesh, Myanmar, Sumatra, Java.
● *C. godefroyi* not recorded outside this area.

NO SPECIMEN

Habitat

In Lao grows in scrub, often in areas which flood every year, at 200–300 m.

Fruiting and flowering months

Flowers ?
Fruit ripe 4–5 (Lao).

Uses

Handicrafts: suitable.
Trade: suitable.
Shoot: edible.

Planting

Trial plantation at Nam Xuang, Lao. First planted for shoot production in 1994, now more than 70 ha planted from Vientiane Province southwards and spreading rapidly. Also planted in NE Thailand. Small cane plantations reported in Bangladesh and Vietnam.

Stem

Stem up to 20 m long and 0.8–1.6 cm diameter.

Sheath 1–2.5 cm diameter.

Sheath green, often striped white when freshly emerged.

Spines needle-like, green with a black tip, 0.5–2 cm long. Bases form downswept crescents.

Ocrea tiny, dry, with no spines or bristles.

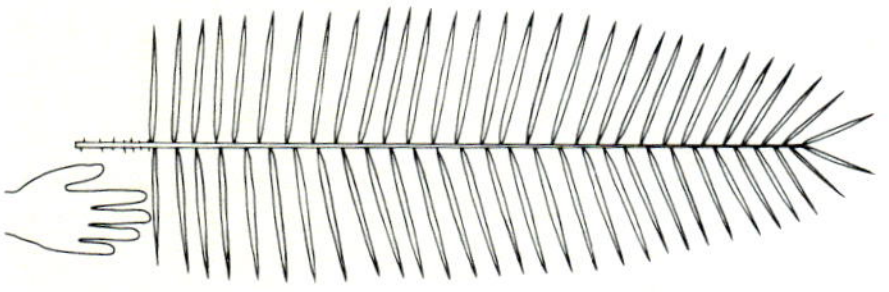

Whole **leaf** 0.6–1 m long.

Petiole of upper leaves 8–15 cm long.

Largest leaflets 10–35 cm × 1–1.5 cm.

Three prominent veins above, all with black bristles. Below mid-vein bristly. Margins bristly. Rachis sometimes has a few claws pointing upwards.

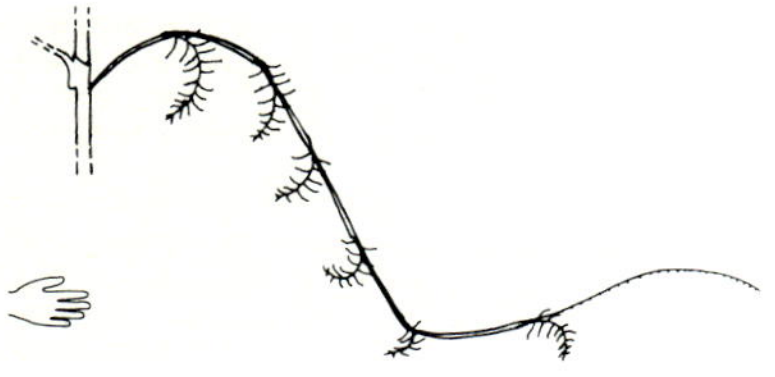

Inflorescence 1.5–2.5 m long plus flagellum. Branches pendulous.

Primary bracts entire, sometimes slightly inflated.

♂ usually 2-branched.
Rachillae up to 10 cm long.

♀ usually 2-branched.
Rachillae up to 7 cm long.

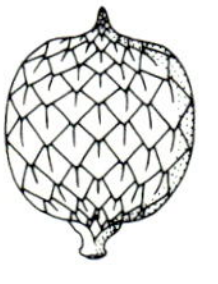

Fruit

1.4 cm long

0.8 cm wide

Scales white or pale yellow with fine dark margins. Perianth has a tubular part.

Seed slightly ruminate.

Compare:

- *C. siamensis* (5): leaflets often interruptedly pinnate with one line of pale bristles above. Sheath spines yellow at base. Male rachillae very short, fruit perianth not tubular.
- *C. walkeri* (5a): ocrea short, densely bristly. Spines not crescent-shaped at the base. Male rachillae very short, fruit perianth slightly tubular.
- *C. guruba* (7): long, papery ocrea, wide, upward-pointing sheath spines, primary bracts open in flat sheets.

● 6a *Calamus godefroyi* (wai nong)

Petiole usually ≤3 cm, the lowest pair of leaflets pointing back across the stem. Leaflets have faint grey indumentum underneath and very few bristles on upper face. Partial inflorescences small, starting deep inside primary bract. Fruit perianth has a tubular part. Fruit ripe 8 (Lao). Grows in marshy areas.

7 *Calamus guruba*

(wai deng)

Climbing Clustering Flagellum Knee conspicuous

An untidy looking species with conspicuous inflorescence bracts.

Look for:

- Sheath with wide and slender spines, all pointing up. Leaflets linear, regular, widely spaced.
- Ocrea conspicuous, papery, longer than 10 cm, twisted.
- Primary bracts open into wide flat sheets.

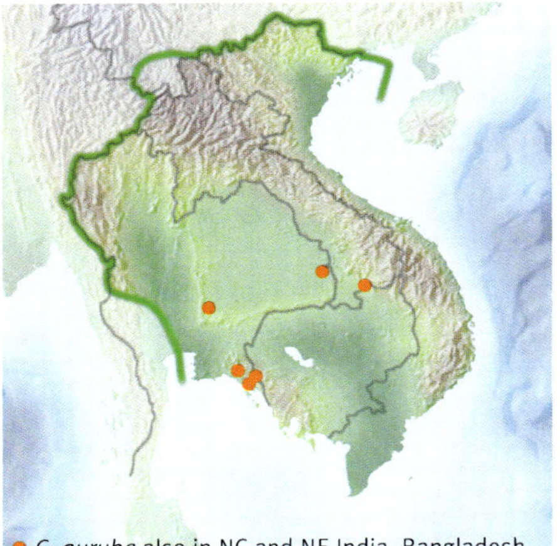

● *C. guruba* also in NC and NE India, Bangladesh, Myanmar, S Thailand, Peninsular Malaysia.

Habitat
In Lao grows in scrub forest at 100 m, often by rivers. In E Thailand recorded up to 750 m.

Fruiting and flowering months
Flowers in bud 11, 12 (Thailand), open 3 (Thailand).
Fruit unripe 12, 3 (Thailand).

Uses
Handicrafts: suitable.
Trade: unknown.
Shoot: edible.

Planting
None known in Lao.

JD 5458

Stem

Stem up to 10 m long and 0.5–1.5 cm diameter.

Sheath 1–2.5 cm diameter.

Sheath dark green.

Spines red-brown, pointing up, some very wide and up to 3.5 cm long, others needle-like, spines near sheath mouth 5-10 cm long.

Ocrea ≥10 cm, red-brown, papery, with no spines or bristles, twisted, falls off easily.

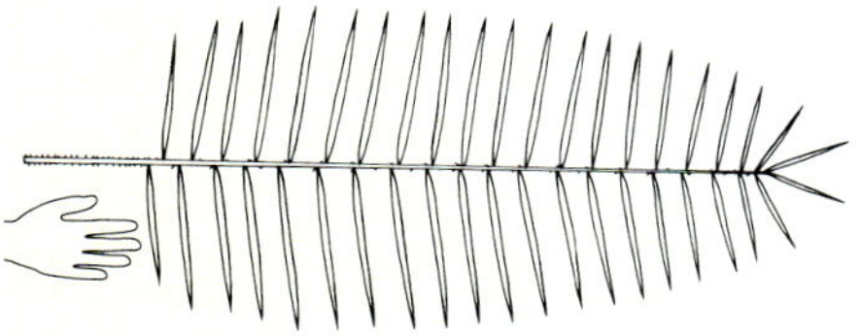

Whole **leaf** 0.6–1.5 m long.

Petiole of upper leaves 10–25 cm long.

Largest leaflets 25–45 cm × 1.7–2.2 cm. Leaflets regular, widely spaced. Above 1-3 prominent veins, some or all bristly. One or more bristly veins below.

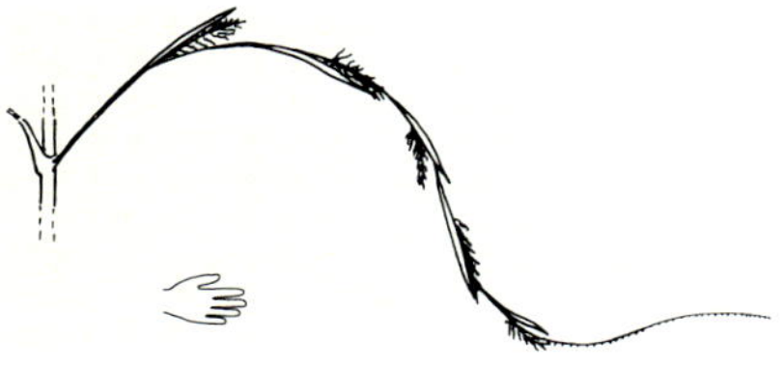

Inflorescence 0.3–2.5 m long plus flagellum. Branches slightly pendulous.

Primary bracts tubular at base, open in a wide sheet at mouth.

♂ usually 3-branched. Rachillae tiny, zigzag.

♀ usually 2-branched. Rachillae zigzag.

Fruit

0.8 cm long

0.8 cm wide

Pale yellow scales with dark margins when dry.

Seed not ruminate.

Compare:

- *C. henryanus* (11): Ocrea tiny, leaflets interrupted pinnate, primary bracts lacerated, without an open limb.
- *C. tenuis* (6): Leaf sheath striped green-white, spines crescent at base with black tip, ocrea tiny, primary bract without an open limb.
- *C. siamensis* (5): Ocrea tiny, leaflets often irregularly pinnate, close together, primary bract without an open limb.
- *C. walkeri* (5a): Ocrea short, covered in bristles, spines green with black tip, primary bract without an open limb.

8 *Calamus poilanei*

(wai thoon, blong thoon, gaparl, probably also wai khom, blong chang)

Climbing Solitary Flagellum Knee conspicuous

The best large Lao cane, heavily harvested. Regenerates poorly because it is solitary-stemmed.

Look for:

- Stem solitary, very robust, leaf sheath often more than 5 cm diameter, with a few thick spines.
- Leaflets regularly arranged, with many bristles on the upper face.
- Petiole margins usually have spines paired, pointing in different directions.

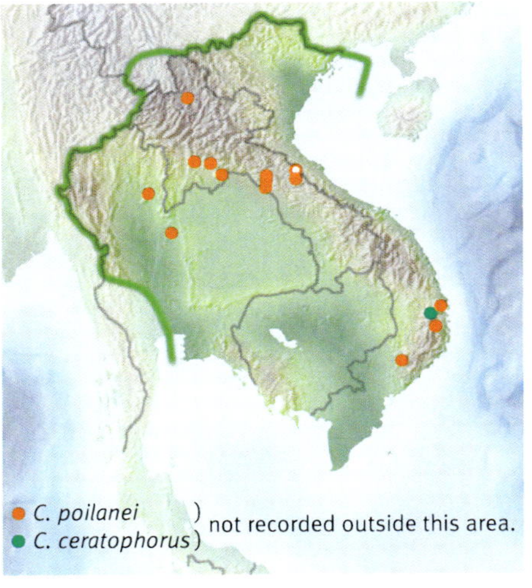

Habitat

In Lao grows in evergreen forest at 300–1300 m. In Thailand at 600–700 m and in S Vietnam at 500 m.

Fruiting and flowering months

Flowers ?

Fruit unripe 4 (Lao), ripe 8 (Lao), 11 (Vietnam).

Uses

Handicrafts: very high quality.

Trade: the most valuable cane in trade in Lao.

Shoot: edible.

Planting

Small trial plantation at Nam Xuang in Lao. High potential, although seedling establishment is currently slow. Reportedly planted in Vietnam.

OL 211

Stem

Stem up to 150 m long and 3–5 cm diameter.

Sheath 2.5–7.5 cm diameter.

Sheath yellowish green, sometimes with patches of dark indumentum.

Spines few, solitary or occasionally in pairs. Broad-based, yellowish-green, black tipped, 1.5–3.5 cm long.

Ocrea tiny, dry, with no spines or bristles.

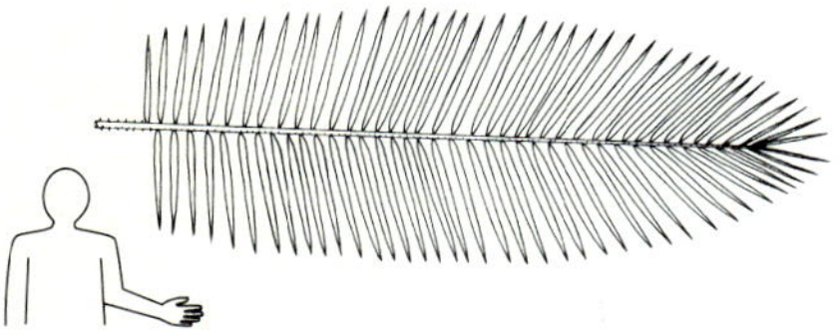

Whole **leaf** 2–4 m long.

Petiole 20–60 cm long. Margin often has pairs of black-tipped spines pointing in different directions.

Largest leaflets 50–80 cm × 2–3 cm.

Above three prominent veins with bristles, below a few bristles on the mid-vein. Margins bristly.

Inflorescence 2–6 m long or more, plus flagellum. Branches quite stiff.

Primary bracts entire, quite tightly sheathing. Armed with stout claws and some short stout spines.

♂ usually 3-branched.

♀ usually 3 -branched. Flower cups not on stalks.

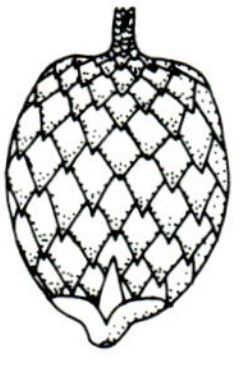

Fruit

2.0 cm long

1.4 cm wide

Scales pale with fine dark margins.

Seed not ruminate.

Compare:

- *C. laoensis* (18), *C. flagellum* (19) and *C. rudentum* (20): all of similar size, but clustering and with sheaths very thickly covered in long flat spines which are black or black-tipped.

- **8a *Calamus ceratophorus***

Climber, quite large, has a cirrus not a flagellum, leaflets regular, sheath spines few and thick, bent down. Leaflets probably lack bristles on both surfaces and margins, petiole spines probably solitary. Female rachillae very distinctive, the flower cups on stalks (see picture). Not known whether solitary or clustering.

9 *Calamus acanthospathus*

(wai hom, blong eur)

Climbing   Solitary or two stems   Flagellum   Knee present, indistinct

A high quality small-diameter cane which may be suitable for cultivation in high-altitude areas.

Look for:

- Leaflets regularly arranged, short, wide, close together and boat-shaped.
- Sheath 1.5–2.5 cm diameter, with many needle-like spines and a short bristly ocrea.
- Grows high in the mountains.

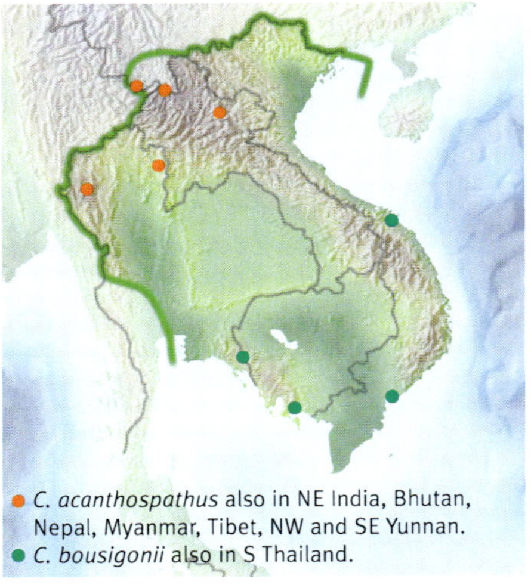

Habitat
In Lao grows in evergreen forest at 1800 m. In Thailand at 1500–1700 m and in S Yunnan at 1600 m.

Fruiting and flowering months
Flowers in bud 2 (Thailand).
Fruit unripe 6 (Lao), ripe 4, 5 (Thailand).

Uses
Handicrafts: high quality
Trade: suitable
Shoot: unknown

Planting
None known in Lao. High potential. Planted for canes in S Yunnan.

OL 231

Stem

Stem up to 30 m long and 0.9 cm diameter.

Sheath 1.5–2.5 cm diameter (up to 5 cm in India).

Sheath dark green with blotches of pale indumentum.

Spines many. Thin, needle-like, 0.5–1 cm long with short fat prickles in between. Break off easily right at the base.

Ocrea about 1 cm long, densely spiny.

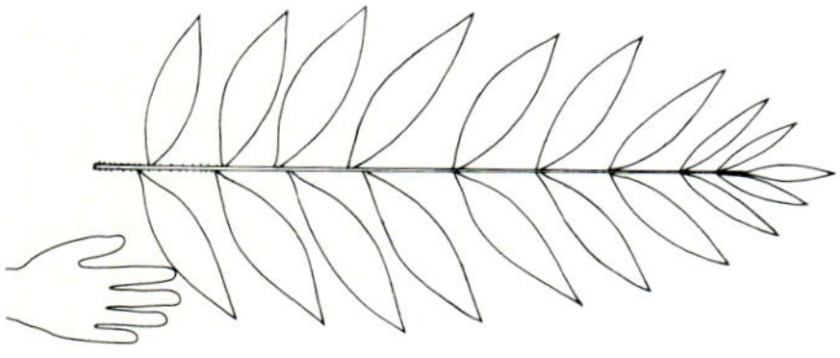

Whole **leaf** 0.5–1 m long.

Petiole of upper leaves 1–3 cm long.

Largest leaflets 18–30 cm × 3.5–7 cm.

Strongly boat-shaped. Above many prominent bristly veins. Below naked. Margins bristly. Veins finish at the tip of the leaflet.

Inflorescence 0.7–1.4 m long plus a short flagellum. Quite stiff.

Primary bracts entire, inflated slightly to cover base of partial inflorescence.

♂ usually 2–3-branched.

♀ usually 2-branched . Wide flower cups make rachillae look a little like squid tentacles.

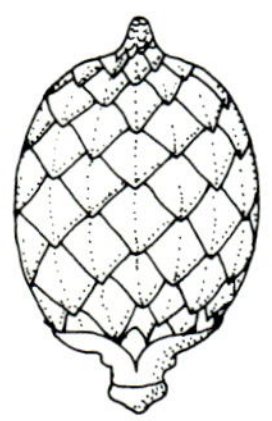

Fruit

2.4 cm long

1.5 cm wide

Scales yellow-green with reddish margins when unripe, dark brown when dry. Channelled when ripe.

Seed ruminate.

Compare:

● **9a *Calamus bousigonii***

Sheath ≤1 cm diameter, petiole 6–20 cm, leaflets few, rhomboid, veins finish at the margins not the tip (see picture), fruit small and yellow when dry. Leaflets like *Korthalsia* but has a flagellum, a knee and no branches. Recorded below 700m in Vietnam and up to 1000 m in Cambodia. Flowers in bud 4, fruit ripe 8 (Vietnam).

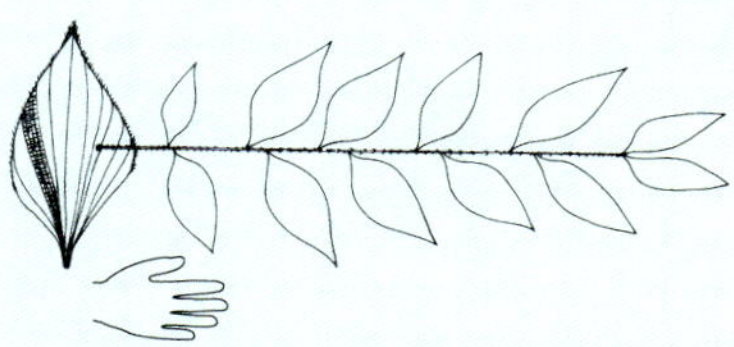

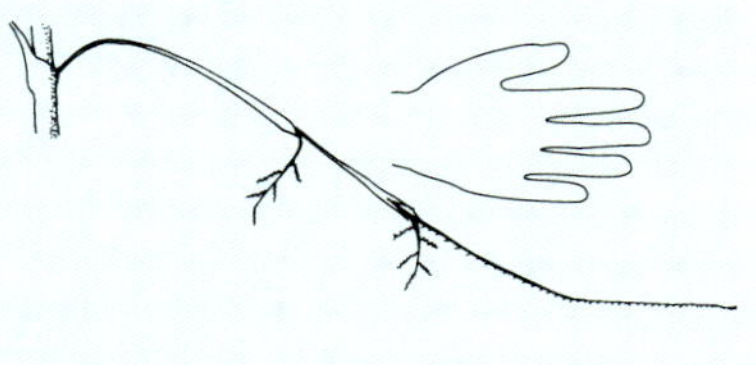

10 *Calamus gracilis*

(wai hom, wai soum, wai tairtair)

Climbing Clustering Flagellum Knee conspicuous

One of the best quality small diameter canes.

Look for:

- Leaf sheath with spines ≤0.5 cm long, swollen, conical at base.
- Leaflets interupted pinnate, dark green. Petiole ≤3 cm long.
- Fruit very distinctive, large and orange when mature and fresh.

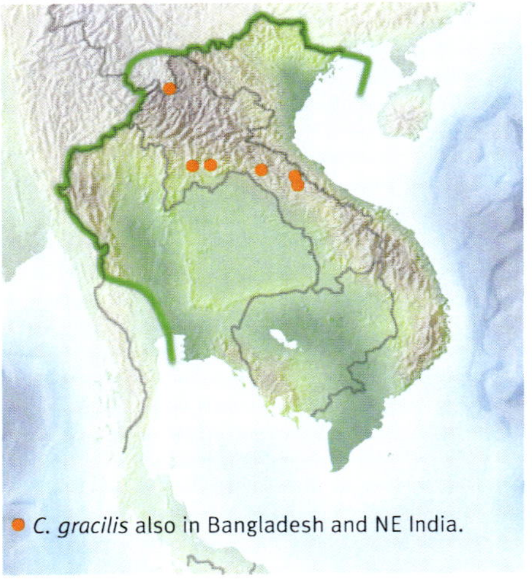

Habitat

In Lao grows in evergreen forest at 300–750 m. Up to 850 m in S Yunnan.

Fruiting and flowering months

Flowers ?

Fruit unripe 5 (S Yunnan), ripe 3 (Lao).

Uses

Handicrafts: high quality.

Trade: one of the preferred species in Lao.

Shoot: edible.

Planting

None known in Lao. High potential.

KP 379

Stem

Stem up to 30 m long and 0.4–1 cm diameter.

Sheath 0.5–2 cm diameter. Sheath dark green with dark indumentum.

Spines green with short black points, swollen and conical at base, slightly pointing up, ≤0.5 cm long.

Ocrea tiny, dry, with no spines or bristles.

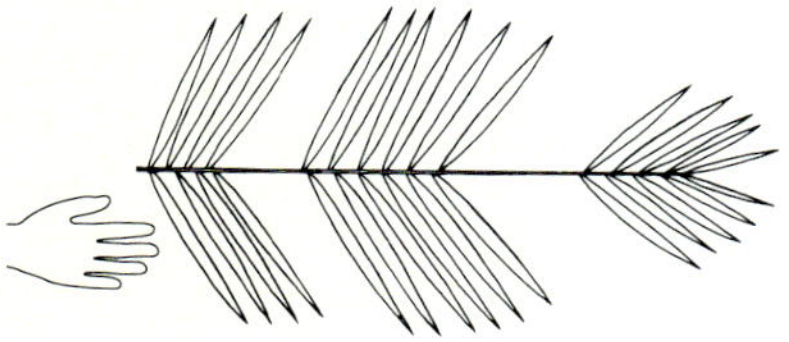

Whole **leaf** 0.3–0.7 m long.

Petiole of upper leaves 0–3 cm long.

Largest leaflets 25–35 cm × 1.5–2.5 cm.

Dark shiny green on both surfaces. 3–5 veins on upper side with bristles, below a few bristles only on middle vein. Margins bristly.

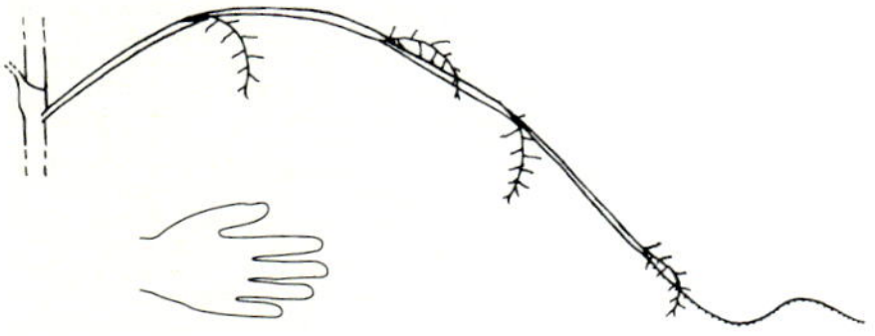

Inflorescence 0.3–0.7 m long plus flagellum. Branches slightly pendulous.

Primary bracts inflated, with short tattered (not lacerated) limb.

♂ (not illustrated) usually 2-branched.

♀ usually 2-branched.

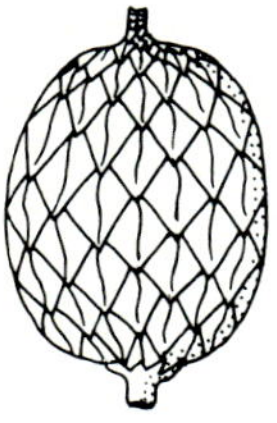

Fruit

2.5 cm long

1.7 cm wide

Scales orange in life, straw-yellow when dry or unripe. Deeply channelled.

Seed ruminate.

Compare:

- *Calamus henryanus* (11): Leaf sheath with many flattened, triangular spines ≥1 cm long, petiole ≥10 cm long, leaflets mid-light green.
- *Calamus siamensis* (5): Leaf sheath spines often ≥2 cm, without swollen, conical bases, petiole ≥8 cm, fruit much smaller.

11 *Calamus henryanus*

(wai namlee, wai hangnou, wai khairp, blong knair)

Climbing Clustering Flagellum Knee conspicuous

Look for:

- Leaf sheath with brownish indument, spines flattened, triangular, 1.5–2 cm long. Ocrea tiny.
- Leaflets interruptedly pinnate, mid-green. Petiole 5-25 cm.
- Rachillae short, pressed against the branch, the tip often curled back.

● *C. henryanus* also in NW and SE Yunnan, SE China.

Habitat

In Lao grows in scrub forest, evergreen forest and evergreen forest on karst at 600–1100 m. In N Thailand at 450–1300 m and in S Yunnan at 1350 m.

Fruiting and flowering months

Flowers in bud 3–6 (Lao), 1–3 (Thailand), 11 (S Yunnan), open 3 (Lao), 12, 3 (Thailand). Fruit unripe 5 (Thailand), 4, 8 (Vietnam).

Uses

Handicrafts: suitable, high quality.
Trade: unsuitable? - stems too short.
Shoot: edible.

Planting

None known in Lao.

KP 380

Stem

Stem up to 8 m long and 0.4–1 cm diameter.

Sheath up to 1.8 cm diameter. Sheath green with red-brownish indumentum.

Spines green with black tips, broad base, flattened, 2 cm long with brown hairy margins. Near sheath mouth sometimes 1–2 erect spines, 4 cm long.

Ocrea tiny, dry, with no spines or bristles.

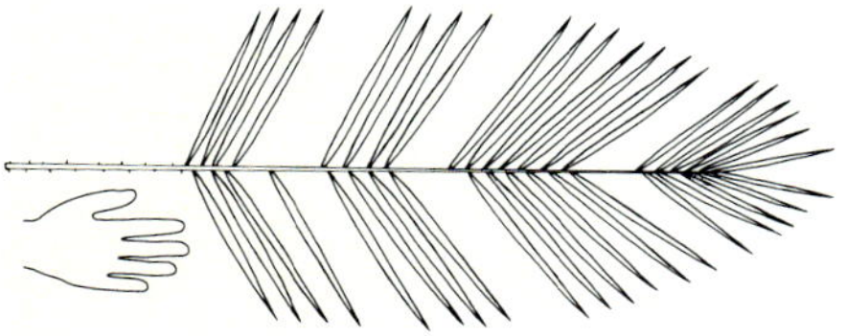

Whole **leaf** 0.4–0.7 m long.
Petiole of upper leaves 5–25 cm long.

Largest leaflets 15–40 cm × 1.3–2 cm.
Above 3–5 prominent veins, all bristly.
Margins bristly.

Inflorescence 0.6–1.5 m long plus flagellum.
Branches pendulous at their tips.

Primary bracts dry and lacerate at the mouth. The terminal rachillae are 2–3 times longer than the others.

Male and female rachillae 0.5–2 cm long (except terminal ones), pressed against the branch, the tip curved back.

♂ usually 2–3-branched.

♀ usually 1–2-branched.

Fruit

1.0 cm long

0.7 cm wide

Scales straw-coloured (when dry) with red margins. Slightly channelled.

Seed not ruminate.

Compare:

- *C. gracilis* (10): Leaf sheath dark green, spines very few, ≤0.5 cm, swollen base. Petiole ≤3 cm, leaflets dark green. Fruit very different.
- *C. guruba* (7): leaflets regularly arranged, ocrea 5–10 cm. Inflorescence very different.
- *C. siamensis* (5): sheath green with little indumentum and narrower spines, leaflets with one line of bristles above, inflorescence well-branched.

12 *Calamus solitarius*

(wai thork, wai yong, wai hakyong, wai savang)

Climbing Solitary Flagellum Knee conspicuous

One of the most important small diameter canes in Lao, common and of high quality.

Look for:

- Leaflets in groups of 2–3 on each side of the rachis, with few or no bristles.
- Last pair of leaflets joined for 20–60% of their length.
- Solitary habit. Mature stems have long inflorescences and partial inflorescences ≥60 cm.

● *C. solitarius* not recorded outside this area.

Habitat

In Lao grows in evergreen forest at 200–600 m. In NE Thailand at 200 m.

Fruiting and flowering months

Flowers ?
Fruit unripe 12 (Lao), 2 (Thailand).

Uses

Handicrafts: high quality.
Trade: one of the preferred species in Lao.
Shoot: edible but very small.

Planting

Small trial plantation at Nam Xuang.
Moderate potential.

OL 240

Stem

Stem up to 50 m long or more and 0.4–1 cm diameter.

Sheath 0.6–1.5 cm diameter.

Sheath dark or pale green with some grey-brown indumentum.

Spines few or many. Green, slender, up to 2 cm long, with slightly swollen base. The roots often form stilts raising the base of the stem off the ground (hence 'wai hakyong') but this does not help to identify it.

Ocrea tiny, dry, with no spines or bristles.

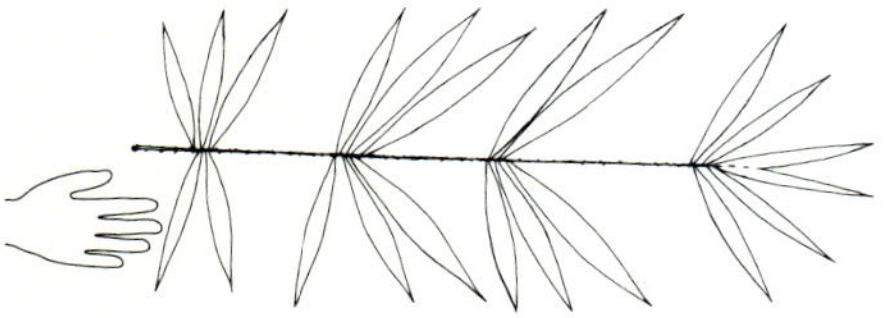

Whole **leaf** 0.5–1 m long.

Petiole of upper leaves 2–10 cm long.

Largest leaflets 15–32 cm × 3.5–4.5 cm.

Three prominent veins above, central one occasionally with a few bristles. Margins slightly or not at all bristly.

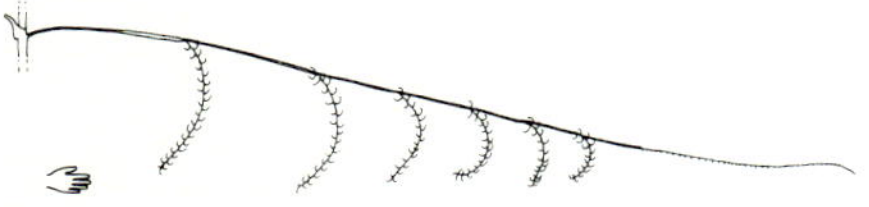

Inflorescence 1.3–5 m plus flagellum.
Pendulous branches.
Primary bracts entire, tightly sheathing.

In both sexes, longest partial inflorescence usually ≥60 cm long and rachillae often curved right back.

♂ usually 3-branched.

♀ usually 2-branched.

Fruit (not illustrated)

≥0.8 cm long

≥0.8 cm wide

Scales pale yellow (when dry) with reddish margins. Not seen ripe. Perianth remains partly tubular.

Seed not ruminate.

Compare:

- *C. tetradactylus* (14): clustering habit, longest partial inflorescences ≤40 cm.
- *C. oligostachys* (13): clustering habit, inflorescences very little branched, partial inflorescences ≤10 cm. Fruit perianth not tubular.
- *C. dioicus* (15a): ocrea bristly, leaflets with three lines of bristles above, last pair not joined, longest partial inflorescence ≤40 cm.
- *C. hypoleucus* (17) and *C. salicifolius* (17a): no climbing organ, inflorescences erect, ≤50 cm long.

13

Calamus oligostachys

(wai kating)

Climbing Clustering Flagellum Knee conspicuous

Look for:

- Leaflets in groups of 2–3 on each side of the rachis, with few or no bristles.
- Last pair of leaflets joined for 20–50% of their length.
- Clustered habit. Long inflorescence, partial inflorescences ≤10 cm, not or hardly branched.

● *C. oligostachys* not recorded outside this area.

Habitat

In Lao in bamboo scrub on silty soil close to a seasonal stream at 150 m. In NE Thailand in evergreen forest at 200 m.

Fruiting and flowering months

Flowers ?
Fruit unripe 12 (Lao), 2 (Thailand).

Uses

Handicrafts: suitable.
Trade: suitable?
Shoot: edible, small.

Planting

None known in Lao.

KP 324

Stem

Stem up to 7 m long and 0.5 cm diameter.

Sheath 0.5–1 cm diameter.

Sheath green.

Spines yellow-brown, 0.2–0.7 cm.

Ocrea tiny, dry, with no spines or bristles.

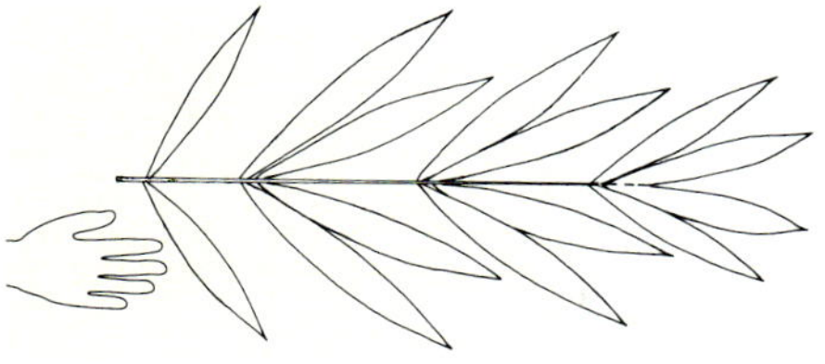

Whole **leaf** 0.4–0.6 m long.

Petiole of upper leaves ≤5 cm long.

Largest leaflets 14–25 cm × 1.8–3.4 cm.

Three prominent veins above, central one occasionally with bristles. Margins smooth or sometimes with a few bristles near the tip.

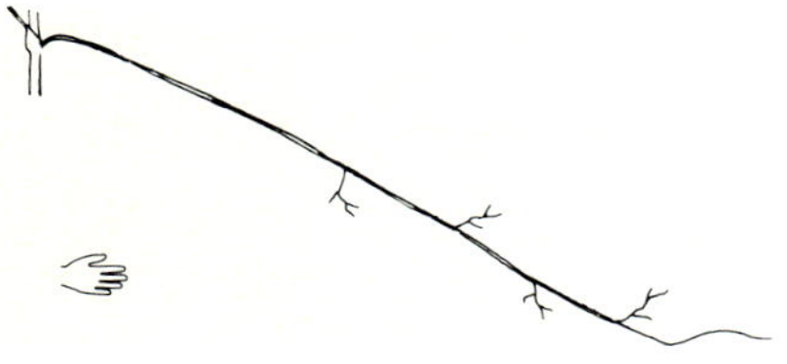

Inflorescence 1.3–2.2 m long plus flagellum. Branches slightly pendulous.

Primary bracts entire and tightly sheathing.

♂ usually 1–2-branched. Each male partial inflorescence ≤11 cm long and has only 0–2 tiny second order branches.

♀ usually 1 -branched. The female rachilla is ≤10 cm long, arises from the axis and has a 1–2 cm stalk with no flowers.

Fruit (not illustrated)

≥0.7 cm long

≥0.7 cm wide

Scales yellow (when dry) with fine dark margins. Fully ripe fruit unknown. Perianth not tubular.

Seed not ruminate.

Compare:

- *C. tetradactylus* (14): partial inflorescences well branched, longest ones ≥20 cm, perianth partly tubular.
- *C. solitarius* (12): solitary habit, longest partial inflorescences ≥60 cm and well branched.
- *C. dioicus* (15a): ocrea bristly, leaflets with three lines of bristles above, last pair not joined, inflorescence 2-branched
- *C. hypoleucus* (17) and *C. salicifolius* (17a): no climbing organ, inflorescences erect, ≤50 cm long.

14 *Calamus tetradactylus*

(wai hangnou, wai hangnou nyai, wai savang, kaceck doikanair, re peu)

Climbing Clustering Flagellum Knee conspicuous

A high quality small-diameter cane widely planted in Vietnam and China.

Look for:

- Slender stem and leaflets usually in groups of 2–3 on a side. Last pair joined 20–40%.
- Leaflets usually with bristly margins.
- Clustering habit. Longest partial inflorescences 10–40 cm, well branched.

● *C. tetradactylus* also in Yunnan and SE China.

Habitat
Evergreen forest and scrub at 100–1000 m.

Fruiting and flowering months
Flowers in bud 8 (Lao), 6, 8 (Thailand), open 8 (Lao). Fruit unripe 1, 5 (Lao), 12 (Thailand), 7–1 (Vietnam). Ripe 5 (Lao), 3 (Cambodia), 4 (Vietnam).

Uses
Handicrafts: suitable, high quality.
Trade: one of the preferred species in Vietnam.
Shoot: unknown.

Planting
None known in Lao. Reported to be commonly cultivated in Vietnam (especially N) for cane production in plantations and home gardens.

KP 399

Stem
Stem up to 6 m long and 0.3–1 cm diameter.

Sheath 0.7–1.8 cm diameter.

Sheath dark green.

Spines yellow-brown with dark brown tips, 0.2–1.5 cm, varying from many to few.

Ocrea tiny, dry, with no spines or bristles (may be 1 cm long when the leaf is still opening).

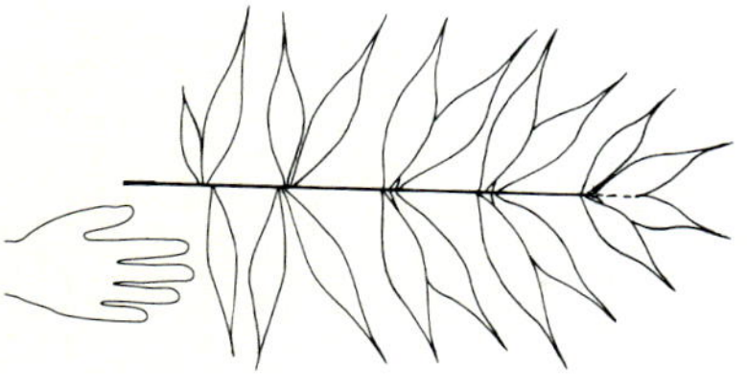

Whole **leaf** 0.2–0.6 m long.

Petiole of upper leaves 3–10 cm long.

Largest leaflets 11–28 cm × 2–6.5 cm.

Grouping of leaflets sometimes quite irregular. Three prominent veins above , the middle one usually naked, sometimes with a few bristles. Margins bristly except in Bolikhamxay Province, Lao.

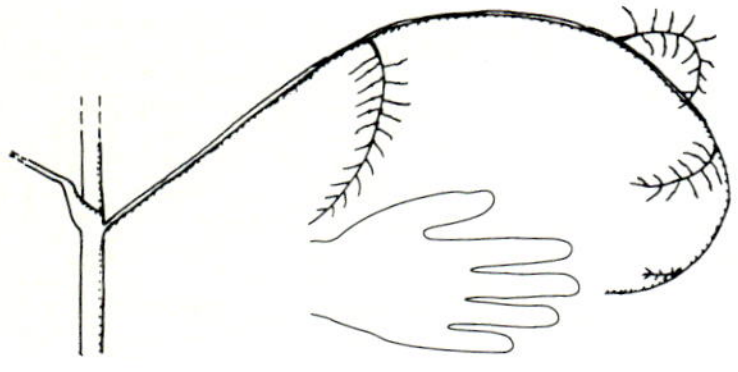

Inflorescence 0.4–1.8 m long plus (usually) a flagellum. Branches pendulous.

Primary bracts entire, either tightly sheathing or slightly inflated near base of inflorescence.

Partial inflorescences of both sexes 10–40 cm long. Male and female rachillae sometimes curved right back.

♂ usually 3-branched.

♀ usually 2-branched.

Fruit

0.9 cm long

0.9 cm wide

Scales pale yellow (when dry) with fine dark margins. Perianth partly tubular.

Seed slightly ruminate.

Compare:

- *C. solitarius* (12): solitary, high climbing habit, longest partial inflorescences ≥60 cm.
- *C. oligostachys* (13): partial inflorescences ≤10 cm, not or very little branched. Fruit perianth not tubular.
- *C. dioicus* (15a): ocrea bristly, leaflets with several lines of bristles above, last pair not joined.

15 *Calamus kingianus*

(wai leum)

Climbing Clustering Flagellum Knee conspicuous

Look for:

- Largest leaflets usually ≥40 cm, irregularly grouped.
- Leaflets have bristly upper surface but no marginal bristles.
- Inflorescence and sheath with a rough surface made up of tiny prickles. Short, bristly ocrea.

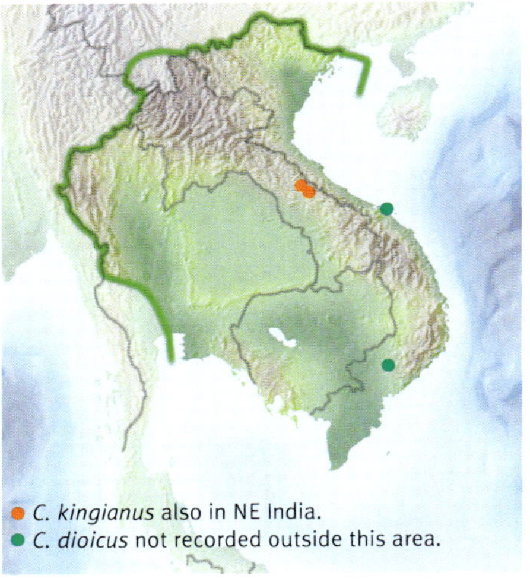

Habitat
In Lao grows in degraded evergreen forest on flat ground on the Nakay Plateau, 500–550 m.

Fruiting and flowering months
Flowers open 3.
Fruit ?

Uses
Handicrafts: suitable.
Trade: unknown.
Shoot: unknown.

Planting
None known in Lao.

AKP 366/367

Stem
Stem up to 7 m long and 0.6–1 cm diameter.

Sheath 1–1.3 cm diameter.

Sheath green, striped with paler indumentum when freshly emerged.

Spines green or yellow-brown, with black tips, 0.2–1 cm, stiff, brittle. Also many tiny prickles giving a rough surface.

Ocrea ≤1 cm, densely bristly.

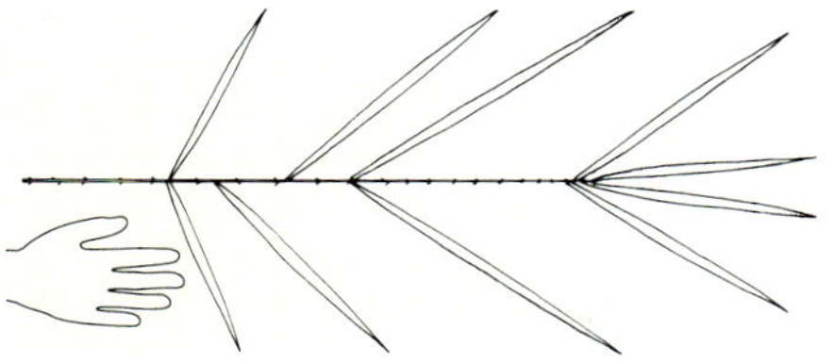

Whole **leaf** 0.4–0.5 m long.

Petiole of upper leaves 15–21 cm long.

Largest leaflets 40–43 cm × 2–3.7 cm.

Three prominent veins above, all naked or slightly bristly. Margins smooth or with 1–2 bristles near tip. Last pair often not joined.

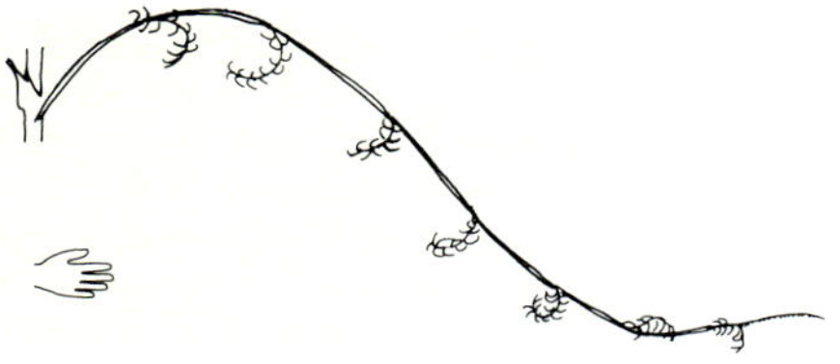

Inflorescence 1.5–3 m long plus flagellum. Branches pendulous.

Primary bracts entire and tightly sheathing. Rough prickly surface.

♂ usually 3-branched. Flowers close together (1.7–1.8 mm spacing) .

♀ usually 2-branched. Rachillae look thick and messy. Flowers close together, bracts wrinkled.

Fruit

Ripe fruit unknown.

Compare:

- *C. solitarius* (12), *C. oligostachys* (13), *C. tetradactylus* (14): leaflets ≤30 cm, usually strongly grouped, inflorescence and sheath not rough, no ocrea, rachillae differ.

● 15a *Calamus dioicus*
Small bristly ocrea, sheath spines soft and flexible, leaflets with bristly margins and three lines of bristles above, last pair of leaflets not joined, inflorescence and sheath not rough, female rachillae smaller and thinner, male flowers at 3.2–3.5 mm intervals. In evergreen forest at 200 m.

16 *Calamus bimaniferus*

(wai hangnou, wai hangnou noi, wai keekai, re itch)

Climbing Clustering Flagellum and cirrus Knee conspicuous

Look for:

- Very slender, with grouped leaflets.
- The only Lao species with a flagellum and a cirrus. Some leaves have a single tiny leaflet instead of a cirrus.

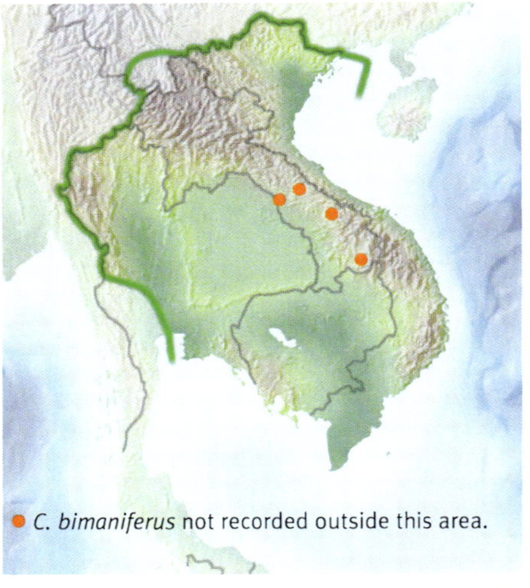

Habitat
In Lao grows in evergreen forest, often degraded areas, at 150–550 m.

Fruiting and flowering months
Flowers ?
Fruit unripe 3, 5.

Uses
Handicrafts: suitable.
Trade: unknown.
Shoot: too small to eat.

Planting
None known in Lao.

KP 369

Stem

Stem up to 3 m long and 0.3–0.4 cm diameter.

Sheath 0.5–1 cm diameter.

Sheath green with some brown indumentum.

Spines often very few, sometimes absent. Green at base, brown-black tips, 0.1–0.7 cm long.

Ocrea tiny, dry, with no spines or bristles.

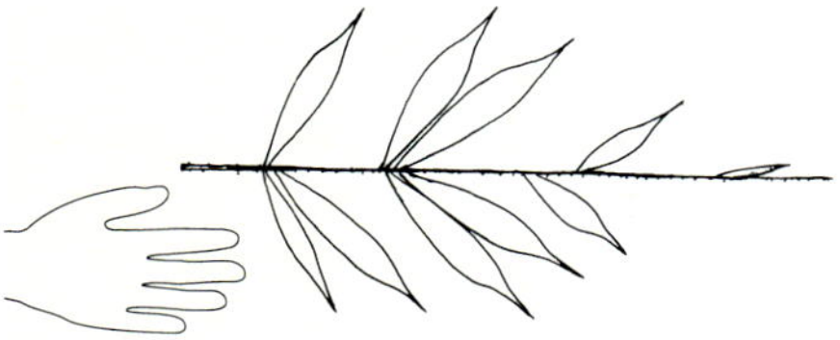

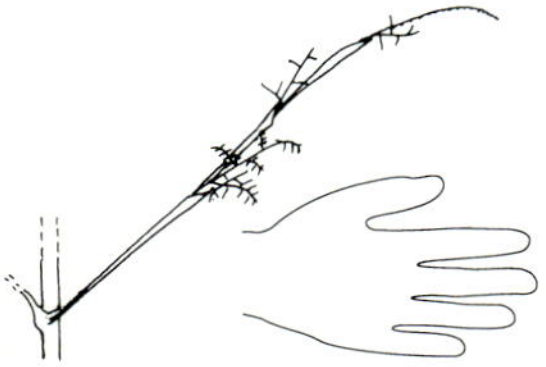

Whole **leaf** 0.3–0.5 m long plus (usually) a cirrus of ≤15 cm. Leaves without a cirrus usually end in a single, very small, clawed leaflet, rarely a pair.

Petiole of upper leaves 8–15 cm long.

Largest leaflets 10–15 cm × 1.5–2.6 cm.

Three prominent veins, naked above and below. Margins also naked, or sometimes with a few small bristles.

Inflorescence 0.2–0.7 m long with no flagellum. Erect.
Primary bracts entire, loosely sheathing.

♂ usually 3-branched.

♀ usually 3-branched.

Fruit (not illustrated)

≥0.8 cm long

≥0.8 cm wide

Scales pale with a dark margin. Ripe fruit not seen.

Seed not ruminate.

Compare:

- *C. solitarius* (12), *C. oligostachys* (13), *C. tetradactylus* (14), *C. kingianus* (15) and *C. dioicus* (15a) all have similar leaves but no cirrus and pendulous inflorescences.
- *C. hypoleucus* (17) and *C. salicifolius* (17a): no climbing organs.

17 *Calamus hypoleucus*

(wai deng, wai kamlao)

Climbing Clustering No climbing organ Knee conspicuous

Look for:

- Leaflets far-apart, irregularly spaced with white undersides and no bristles above.
- Short climbing stems without climbing organs.
- Short infloresences, the primary bracts opening into limbs to cover the partial inflorescences.

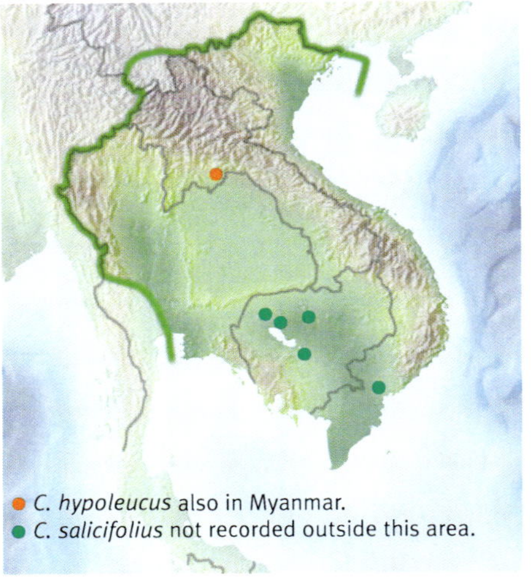

Habitat
In Lao grows in scrub and bamboo at 150 m.

Fruiting and flowering months
Flowers in bud 12 (Lao).
Fruit ?

Uses
Handicrafts: suitable, but quite short.
Trade: not suitable.
Shoot: edible, but small.

Planting
None known in Lao.

KP 319

Stem

Stem up to 3 m long and 0.3–0.5 cm diameter.

Sheath 0.7–1 cm diameter.

Sheath pale green.

Spines yellow-brown with black tips. Mostly 0.2–1.4 cm long, up to 2.5 cm near sheath mouth.

Ocrea tiny, dry, with no spines or bristles.

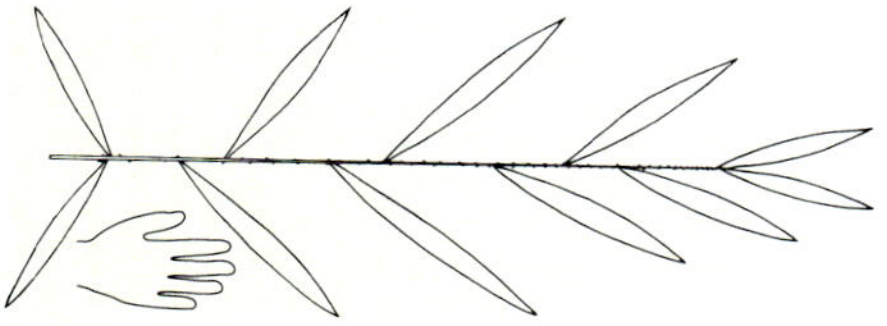

Whole **leaf** 0.4–0.8 m long.

Petiole of upper leaves 3–10 cm long.

Largest leaflets 15–30 cm × 2–4 cm.

Underside with grey-white indumentum. 5–7 prominent veins above, all naked. Margins bristly.

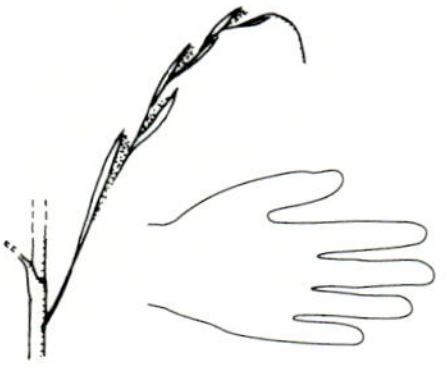

Inflorescence 0.2–0.4 m long with no flagellum or a short stiff one. Erect.

Primary bracts open out to form flat limbs which are longer than the partial inflorescence and red inside.

♂ usually 3-branched.

♀ usually 2-branched.

Fruit

Fruit and seed unknown.

Compare:

- *C. guruba* (7): primary bracts similar but is a high climber with flagellum and regularly spaced leaflets.

• 17a *Calamus salicifolius*

Ocrea 3–5 mm, bristly. Leaflets usually spiny, grouped and pointing in different directions like a tiny *C. viminalis* (4), last leaflets 1–4 cm long. Primary bracts cover ≤1/2 of the partial inflorescence. Flowers in bud 1, 7, open 7, fruit unripe 2 (Cambodia).

18 *Calamus laoensis*

(wai leum, wai katok, wai wa, wai keyomee)

Climbing Clustering Flagellum Knee present, indistinct

A large, newly discovered species only known from central Lao.

Look for:

- Robust habit, the leaf sheath covered in many long flat spines, pale with dark tips.
- Leaflets in groups of 2–5 on each side of the rachis, pointing in all directions.
- Leaflets with white indumentum underneath.

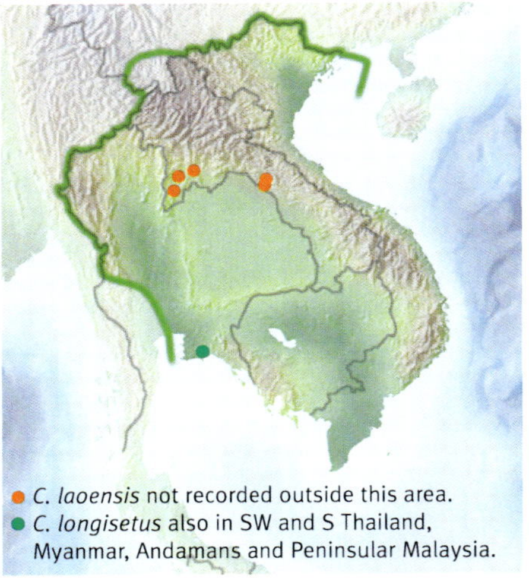

Habitat
In Lao grows in evergreen forest, including forest near karst, at 200–350 m.

Fruiting and flowering months
Flowers in bud 5.
Fruit appears nearly ripe 5. May take ≥1 year to ripen.

Uses
Handicrafts: suitable.
Trade: not preferred, but can be sold.
Shoot: edible (some say it is too bitter to eat).
Leaves can be used as thatch.

Planting
None known in Lao. May have good potential.

KP 352

Stem
Stem up to 40 m long and 2–2.5 cm diameter.

Sheath 3.5–5 cm diameter.

Sheath light green-grey or pale yellow green.

Spines very many. Some long and flat (up to 6 cm), greeny-yellow bases, pale straw colour with dark tips. Also many tiny black needles.

Ocrea small, formed by the brittle margin of the sheath, covered with spines like the sheath.

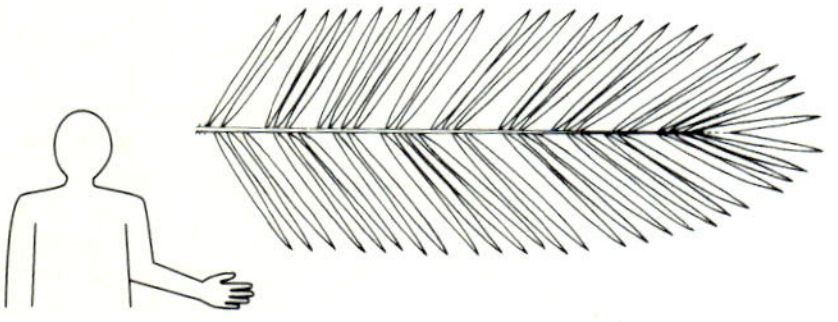

Whole **leaf** 1.7–2.5 m long.

Petiole on upper leaves 5–10 cm long (up to 40 cm on some middle leaves).

Largest leaflets 35–65 cm × 3.2–4 cm.

Underside of leaflets with white indumentum, although this may have washed off on old leaves. One mid-vein prominent, naked on both surfaces. Margins bristly. More strongly grouped on lower leaves.

Inflorescence 2.5–6 m long plus flagellum. Branches pendulous.

Primary bracts splitting at mouth, tightly sheathing.

♂ usually 3–4-branched. Rachilla bracts up to 4 mm wide.

♀ usually 2-branched. Rachilla bracts up to 6 mm wide.

Fruit

2.0 cm when nearly ripe

1.2 cm wide

Scales orange when dry with dark brown submarginal band, deeply channelled.

Seed not ruminate.

Compare:

- *C. flagellum* (19), *C. rudentum* (20): leaflets regularly spaced, not white below.

• 18a *Calamus longisetus*

Leaflets grouped but not white below and have long black bristles on central vein. Rachilla bracts 9–12 mm wide in both sexes. Fruit 3.0 × 2.0 cm, unchannelled and with beautiful black, brown and cream stripes. Cultivated in South Thailand.

19 *Calamus flagellum*

(*wai lao, wai mon, wai namleuang, wai thoon, blong poul,* probably also *wai khanyai, wai long*)

Climbing Clustering Flagellum Knee present, indistinct

Look for:

- Robust habit, the leaf sheath covered in many long flat spines, pale with dark tips.
- Leaflets long, regularly arranged.
- Newly emerged leaf sheaths with a pair of bristly 'ears' which soon fall off.

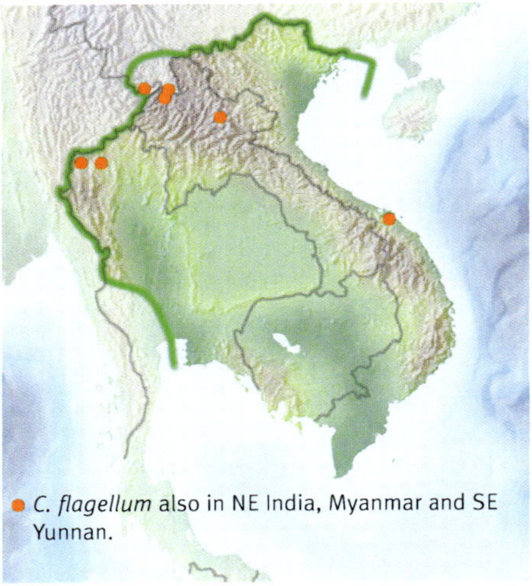

● *C. flagellum* also in NE India, Myanmar and SE Yunnan.

Habitat

In Lao grows in evergreen forest, sometimes on karst, at 800–950 m. Recorded at 1350 m in Vietnam and 1400–1500 m in Thailand.

Fruiting and flowering months

Flowers in bud 5 (Lao), 1 (Vietnam). Open 5 (Thailand).
Fruit ripe, 6 (Thailand). May take ≥1 year to ripen.

Uses

Handicrafts: suitable.
Trade: suitable? (in India, brittle and useless).
Shoot: edible.

Planting

None known in Lao.

OL 234

Stem

Stem 30 m long or more, and 1.5–3 cm diameter.

Sheath 2.5–5 cm diameter.

Sheath greenish-yellow with reddish or brown indumentum amongst spines.

Spines very many. Some long and flat (up to 6 cm), pale straw colour with extensive dark tips. Also many tiny black needles mixed in.

Ocrea over 5 cm, covered in bristles, in the form of an 'ear' each side of the sheath mouth. Visible on the newest sheaths but falls off quite quickly.

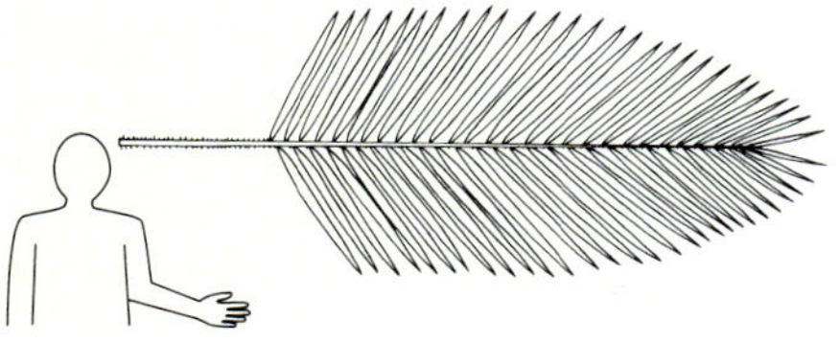

Whole **leaf** 2–2.5 m long.

Petiole of upper leaves 30–70 cm long.

Largest leaflets 60–70 cm × 3.5–4.5 cm.

Mid veins prominent and bearing one line of long or short black bristles on each surface. Margins bristly. The variety *furvifurfuraceus* has dark indumentum on the sheath, petiole and rachis and some lower leaflets grouped.

Inflorescence 4–6 m long plus flagellum. Branches pendulous.

Primary bracts lacerate at the mouth, slightly inflated.

♂ usually 3-branched. Flowers 4.5–6 (rarely 12) mm apart in each row, rachilla bracts 4–5 (rarely 9) mm wide, flowers 8–9 mm long.

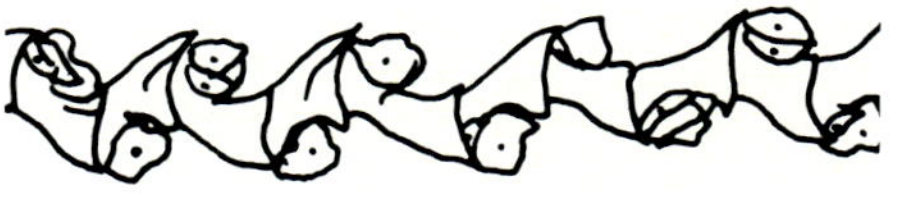

♀ usually 2-branched. Flowers usually 15 mm apart in each row, rachilla bracts 7–8 mm wide, flowers 10 mm long.

Fruit

2.7–3.0 cm long

1.8 cm wide

Big, with channelled scales. Orange-brown when dry.

Seed ruminate.

Compare:

- *C. rudentum* (20): has the leaf sheath almost open on one side, no 'ears', a comb of long petiole spines embracing the stem, thinner rachillae and smaller yellow fruit with the seed not ruminate.
- *C. laoensis* (18): has no 'ears' and grouped leaflets which are white underneath.
- *C. rhabdocladus* (21): has interruptedly pinnate leaflets rarely bigger than 40 × 2 cm, no 'ears' and short male rachillae.
- *C. erectus* (3): has'ears', but is erect with no climbing organ and rings of yellow spines around the petiole.

20 *Calamus rudentum*

(wai boun, boun, probably also wai tabong, boun dam, boun khao)

Climbing Clustering Flagellum Knee present, indistinct

Look for:

- Robust, the leaf sheath covered in many long flat spines, pale with dark tips.
- Leaf sheath deeply split on one side. Long petiole spines embrace the stem.
- Leaflets large, regularly arranged. Petiole with a smooth channel in the upper surface.

● *C. rudentum* also in S and SW Thailand.

NO SPECIMEN

Habitat

In Lao in scrub near a village at 200 m. Probably also in evergreen forest in S Lao. In N and SE Thailand in evergreen forest at 100–400 m.

Fruiting and flowering months

Flowers in bud 3 (Lao), 3, 5 (Thailand), open 3 (Lao). Fruit unripe in all months (Lao), 1, 5–7 (Thailand), 6, 12 (Vietnam), ripe 6–7 (Lao).

Uses

Handicrafts: high quality.
Trade: suitable.
Shoot: edible.
Fruits sometimes sold as food.

Planting

Small trial plantings at Nam Xuang and Ban Nongpennua, Vientiane Prefecture. High potential. Commercial planting reported in Vietnam.

Stem

Stem up to 30 m long and 2.5–3.5 cm diameter.

Sheath 4–7 cm diameter.

Sheath light green or yellowish green.

Spines many. Some long and flat (up to 6 cm), pale straw colour with dark tips. Also many tiny black needles mixed in. At the base of the petiole combs of spines up to 10 cm long are swept back across the stem.

Ocrea tiny.

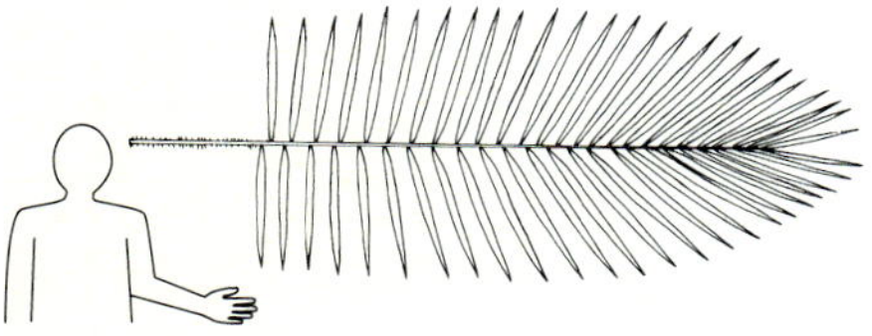

Whole **leaf** 1.8–3.8 m long.

Petiole of upper leaves 20–60 cm long, with a smooth channel on the upper face.

Largest leaflets 45–65 cm × 1.5–2.5 cm.

Mid-veins prominent, with one line of black bristles ≥1 cm long above and below. Margins bristly.

Inflorescence 3.5–6.5 m long plus flagellum. Branches pendulous.

Primary bracts lacerate at the mouth. Slightly inflated.

♂ usually 2–3-branched. Flowers 3–3.5 mm apart in each row, rachilla bracts 2–2.5 mm wide, flowers 6 mm long.

♀ usually 2-branched. Flowers 7.5–9 mm apart in each row, rachilla bracts 4.5 mm wide.

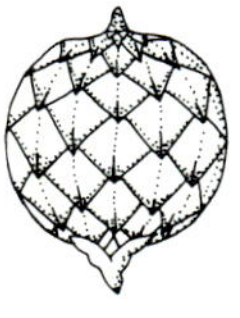

Fruit

1.8 cm long

1.2 cm wide

Yellow when ripe with fine dark margins, scales channelled.

Seed not ruminate.

Compare:

- *C. flagellum* (19): has bigger leaflets, two bristly 'ears' at leaf sheath mouth, sheath not split, lacks very long petiole spines, rachillae of each sex much bigger, seed ruminate.
- *C. laoensis* (18): leaflets grouped, white underneath.
- *C. rhabdocladus* (21): top of petiole flat and spiny, interruptedly pinnate leaflets and very different male rachillae and fruits.

21 *Calamus rhabdocladus*

(wai wan, boun wan, wai bounwan, wai bounyong, blong salay)

Climbing Clustering Flagellum No knee

Look for:

- Leaf sheath covered in many long, flat, jet-black spines, petiole spiny on upper surface.
- Leaflets often interruptedly pinnate, usually ≤40 × 2 cm, three lines of bristles above.
- Male rachillae much smaller than other species with similar leaf sheaths.

● *C. rhabdocladus* also in NW and SE Yunnan, SE China.

Habitat

In Lao grows in evergreen forest at 500–800 m. Also survives well in scrub. Recorded at 1000–1600 m in Vietnam.

Fruiting and flowering months

Flowers in bud 5 (Lao), open 4, 5 (Vietnam). Fruit unripe 10, 1, 2 (Vietnam).

Uses

Handicrafts: suitable.
Trade: suitable.
Shoot: edible, sweet.

Planting

None known in Lao. May have good potential.

KP 371

Stem

Stem up to 40 m long and up to 2.5 cm diameter.

Sheath 4–6 cm diameter.

Sheath green or yellow with reddish-brown indumentum.

Spines very many, short and long, up to 4 cm, black, all in neat rows. At sheath mouth can be ≥10 cm. Petiole spines may be pale tipped.

Ocrea absent or present as a small bristly tag on top of the base of the petiole.

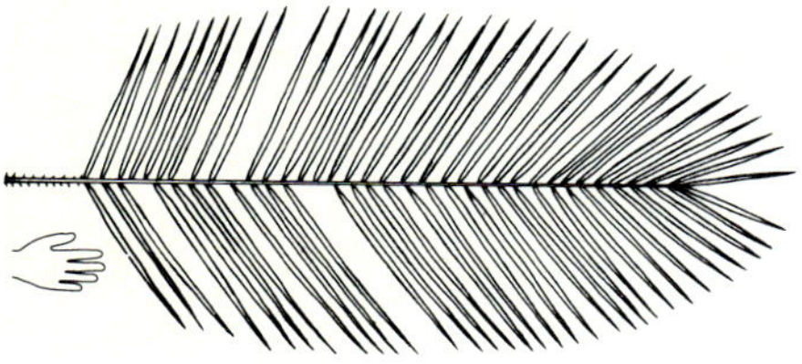

Whole **leaf** 1–2 m long.

Petiole of upper leaves 20–40 cm long, flat and very spiny on upper face.

Largest leaflets 30–49 cm × 1.8–2 cm.

Leaflets regular or interruptedly pinnate, especially at the base of the rachis. Leaflets usually ≤40 cm. Three prominent veins above, all with bristles ≤0.5 cm long. Below a few bristles. Margins bristly.

Inflorescence 2–4 m long plus flagellum. Branches pendulous.

Primary bracts entire, closely sheathing.

♂ usually 3-branched. Rachillae less than 2 cm long, strongly curved.

♀ usually 2-branched. Rachillae up to 8 or rarely 15 cm long.

Fruit

1.4 cm long

0.8 cm wide

Many small, slightly channelled scales, yellow with red-brown margins.

Seed not ruminate.

Compare:

- *C. flagellum* (19): leaflets ≥50 cm, regular with only one line of bristles, leaf sheath with bristly 'ears', petiole with a deep, smooth channel.
- *C. rudentum* (20): leaflets regular, sheath deeply split opposite petiole, spines pale-based male rachillae longer, petiole with a deep, smooth channel.
- *C. laoensis* (18): leaflets grouped, white underneath, petiole with a deep, smooth channel.

22 *Calamus palustris*

(wai hangnou, wai namleuang, wai khairt, wai kiyow, wai hom, wai namhang, wai tiukeng, wai savang, wai sard, wai kanebouang, re tair)

Climbing Clustering Cirrus Knee conspicuous

Widespread and variable. The variety *cochinchinensis* occurs in Indochina.

Look for:

- Cirrate leaves, knee and strongly grouped, boat-shaped leaflets shorter than 40 cm
- Leaf sheath up to 4.0 cm diameter, usually 0.7–2.5 cm.
- Fruit ≤1.2 cm long, with no pedicel, unchannelled scales and a red 'nipple' when fresh.

● *C. palustris* (other varieties) also in Myanmar, S Thailand, Peninsular Malaysia, Andaman and Nicobar Islands.

Habitat

In Lao grows in evergreen forest at 100–650 m. In N and E Thailand up to 1300 m.

Fruiting and flowering months

Flowers in bud 12, 3 (Thailand), open 2 (Lao), 2, 11, 12 (Thailand).
Fruit unripe 2, 5 (Lao), 12, 1, 3 (Thailand), 4 (Vietnam), ripe 4–6 (Thailand).

Uses

Handicrafts: good quality.
Trade: one of the preferred species in Lao.
Shoot: edible.

Planting

Small trial plantation at Nam Xuang, Lao. High potential. Also reportedly grown in Vietnam.

KP 318

Stem

Stem up to 30 m long and 0.3–1.5 cm diameter.

Sheath 0.7–3 (rarely 4) cm diameter.

Sheath green, sometimes striped with white indumentum.

Flowering sheaths usually ≥1.3 cm diameter.

Spines yellow with long red tips, giving a black tip when dry.

Longest spines sometimes 3 cm, usually less. Very variable in number, ranging from a few long spines to many long and small spines mixed.

Ocrea tiny, dry, with no spines or bristles.

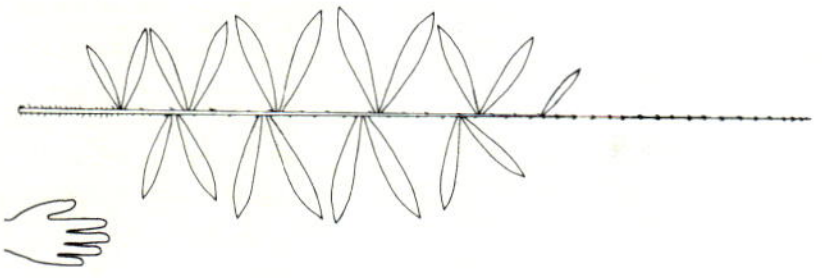

Whole **leaf** 0.65–1.9 m long plus a cirrus of 0.4–1 m.

Petiole of upper leaves 5–30 cm long.

Largest leaflets 17–40 cm × 2.5–7.5 cm.

Leaflets usually ≤35 × 5.5 cm, strongly boat-shaped, a few veins prominent above, all naked. Margins spiny.

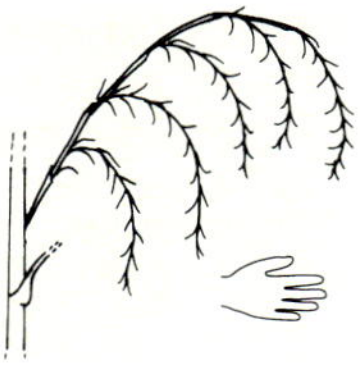

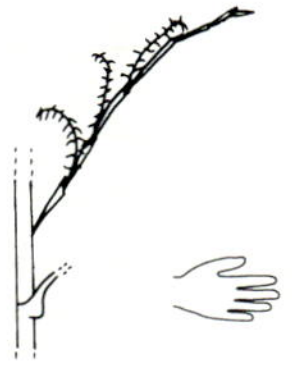

Inflorescence 0.2–1.4 m long with no flagellum. Erect or drooping when longer. Male usually 0.6–1 m, female 0.2–0.5 m.

Primary bracts entire or with neatly defined decaying tip. Slightly inflated, well armed.

♂ usually 3-branched.

♀ usually 2-branched.

Fruit

0.7–1 .2 cm long

0.7–1 cm wide

Scales pale, with red margins and no channels. Fruit tip has a red 'nipple' when fresh.

Perianth has no tubular part. Seed not ruminate.

Compare:

- Other *Calamus* species with a cirrus (23–25): leaflets usually ≥35 cm, sheath usually ≥3 cm diameter, fruit ≥1.6 cm long with tubular perianth and channelled scales. *C. wailong* has regular leaflets, *C. platyacanthus* has very long spines, *C. nambariensis* grows at high altitudes.
- *C. tetradactylus* and similar species (12–15a): similar grouped leaflets, but they have a flagellum, not a cirrus.
- *C. bimaniferus* (16): cirrus smaller (≤15 cm) and a weak flagellum also present.

23 *Calamus nambariensis*

(wai nwn, wai niuw, wai nokkhor, wai namleuang, kateng blor)

Climbing Clustering Cirrus Knee conspicuous

A variable, high altitude species with several easily confused relatives.

Look for:

- Large stem and leaves with cirrus. Leaflets usually grouped, may be regular on some leaves.
- Sheath spines usually a mixture of short and long ones up to 4 cm long.
- Fruit very distinctive, ≥2.5 cm long with deeply grooved scales and a tubular perianth.

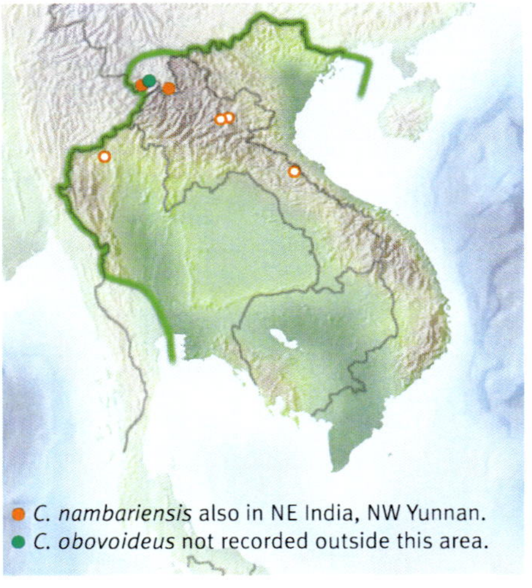

Habitat

Probable records in Lao from two areas of evergreen forest at 1400–1800 m. In S Yunnan at about 1600 m.

Fruiting and flowering months

Flowers in bud 3 (Lao).
Fruit unripe 6 (Lao), ripe 12 (S Yunnan).

Uses

Handicrafts: high quality.
Trade: a preferred species in China.
Shoot: unknown.

Planting

None known in Lao. Several government sponsored plantations in S Yunnan for cane production.

Stem

Stem up to 20 m long and 1–3 cm diameter.

Sheath 1.5–6 cm diameter.

Sheath dull green. Specimens often mossy because it grows at high altitude.

Spines dull yellow-green with dull reddish tips. A mixture of long and short ones, the longest ≤4 cm.

Ocrea tiny, dry, with no spines or bristles.

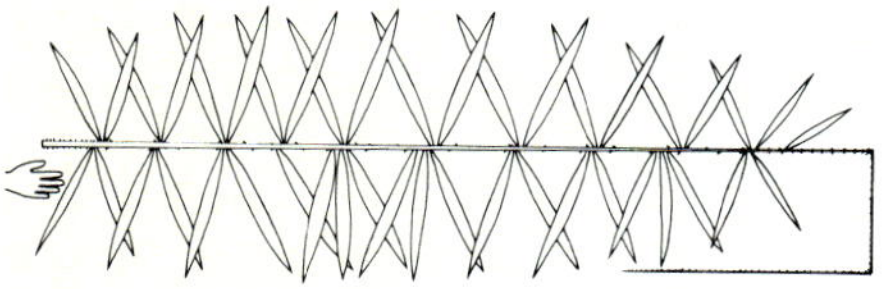

Whole **leaf** 1.5–2.5 m long plus a cirrus about 1 m long.

Petiole of upper leaves 8–20 cm long.

Largest leaflets 40–55 cm × 2.5–7 cm.

Leaflets often grouped, sometimes regular at the base of the leaf or all along it. Width can vary greatly. Several prominent veins above, usually naked or with a few scattered spines. Margins bristly.

Inflorescence 0.65–1.3 m long with no flagellum, erect or, if long, slightly drooping.

Primary bracts entire or with a short, neat, decaying tip. Slightly inflated. Well armed.

♂ usually 3-branched.

♀ usually 2-branched.

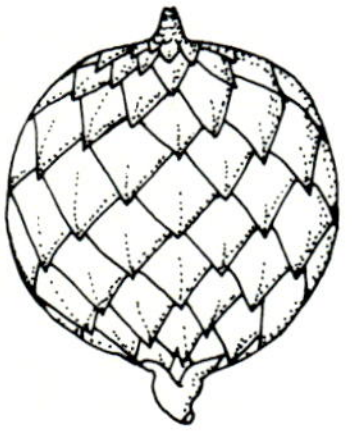

Fruit

2.5 cm long

2.0 cm wide

Scales dull yellowish in life with deeply channelled scales and the perianth partly tubular.

Seed slightly ruminate.

Compare:

- *C. wailong* (25): leaflets always regular, lower sheaths often with very few spines or many very small spines. Ripe fruit ≤2 cm long.
- *C. palustris* (22): leaflets always grouped, sheath usually ≤3 cm diameter, leaflets ≤40 cm, ripe fruit ≤1.2 cm long, unchannelled, fruit perianth not tubular.
- *C. platyacanthus* (24): leaflets always grouped, sheaths spines mainly ≥5 cm. Ripe fruit ≤2 cm long.

● **23a *Calamus obovoideus***
Fruit long, not globose, 3.4 × 2.2 cm, rachis without claws on lower surface.

24 *Calamus platyacanthus*

(wai namleuang, wai leuang)

Climbing Clustering Cirrus Knee conspicuous

A poorly known, variable species easily confused with *C. wailong* and *C. nambariensis*.

Look for:

- Large stem and leaves with cirrus. Leaflets strongly grouped.
- Sheath spines very long (5–9 cm), pointing down, with hardly any small spines amongst them.
- Fruit ≤2.2 cm long, with deeply grooved scales and a partly tubular perianth.

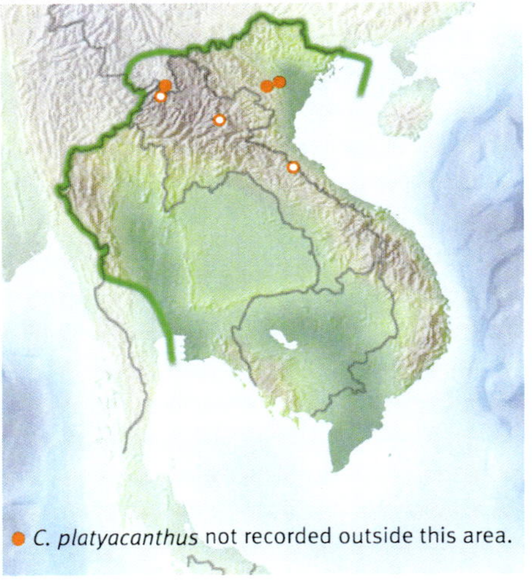

Habitat
Probable records in Lao from evergreen forest at 750–900 m. In Vietnam at 500–700 m.

Fruiting and flowering months
Flowers ?
Fruit unripe 6 (Lao), ripe 8, 11 (Vietnam).

Uses
Handicrafts: suitable, high quality?
Trade: suitable.
Shoot: edible.

Planting
None known in Lao but has high potential. Reportedly a commonly planted species in Vietnam.

BX 01 (left) + KP 387 (right)

Stem

Stem at least 30 m long and 1–2.5 cm diameter.

Sheath 2–6 cm diameter.

Sheath green or yellow-green.

Spines dull yellow. Mainly long (often ≥5 cm or even 9 cm) and pointing down, with no or very few small spines amongst them.

Ocrea tiny, dry, with no spines or bristles.

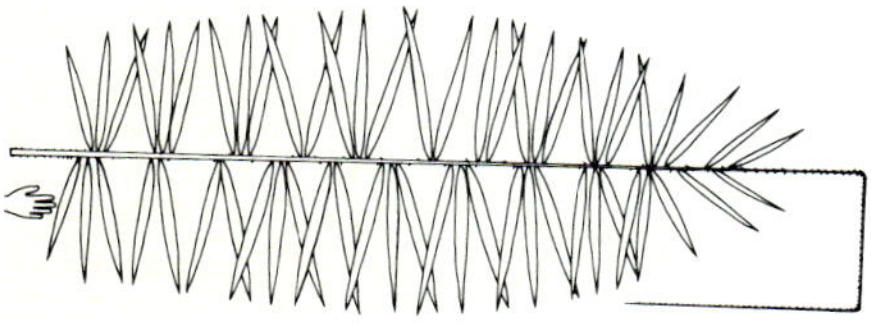

Whole **leaf** 2–3 m long plus a cirrus up to 1.5 m.

Petiole of upper leaves 3–20 cm long.

Largest leaflets 30–50 cm × 4–7 cm.

Width can vary greatly. Several prominent veins above, usually naked or with a few scattered spines. Margins bristly.

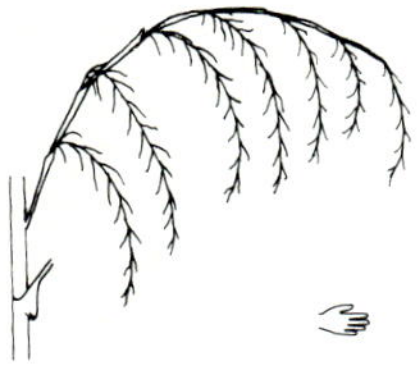

Inflorescence at least 1.5 m long with no flagellum. Erect or slightly drooping when long.

Primary bracts entire or with a neat dry margin. Slightly inflated, armed.

♂ usually 3-branched.

♀ usually 2-branched.

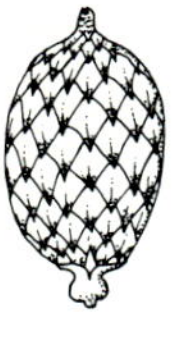

Fruit

1.8–2.2 cm long

1.2 cm wide

Scales deeply channelled with a faint dark margin. Perianth partly tubular.

Seed slightly ruminate.

Compare:

- *C. nambariensis* (23) and *C. obovoideus* (23a): some leaflets regular, a mixture of small and large sheath spines (up to 4 cm) and fruit ≥2.5 cm long.
- *C. palustris* (22): sheath ≤3 cm diameter, leaflets ≤40 cm, fruit ≤1.2 cm long with no grooves and the perianth not tubular.
- *C. wailong* (25): leaflets regular, longest spines ≤4 cm, young sheaths often have no big spines, only tiny ones ≤0.5 cm.

25 *Calamus wailong*

(wai nokkhor, wai khor, wai lai, wai namhang, wai namleuang, wai khairt, kloong, kateng koday)

Climbing Clustering Cirrus Knee conspicuous

Very variable and closely related to both *C. platyacanthus* and *C. nambariensis*.

Look for:

- Large stem and leaves with cirrus. Leaflets regular except on seedling leaves.
- Lower sheaths often only have tiny spines. When present, long spines ≤3 cm.
- Fruit ≤2 cm long with deeply grooved scales and a pedicel.

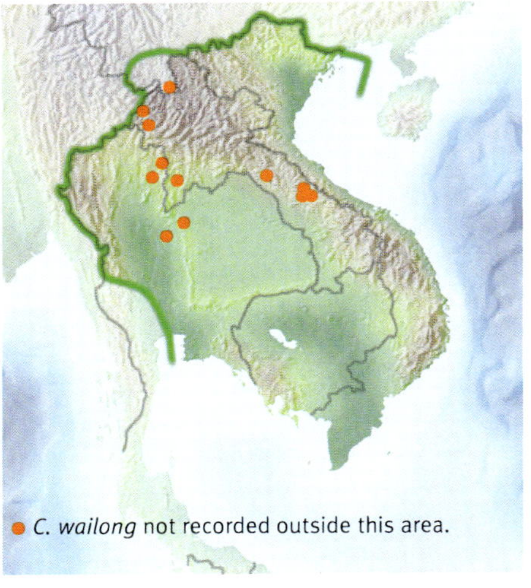

● *C. wailong* not recorded outside this area.

Habitat

In Lao in evergreen forest at 350–600 m and in Thailand at 600–1000 m. In S Yunnan at 600–950 m.

Fruiting and flowering months

Flowers in bud 1, 3 (Lao, Thailand).
Fruit reported ripe 8–11, (Lao), ripe 11 (S Yunnan).

Uses

Handicrafts: suitable, high quality.
Trade: suitable.
Shoot: edible.

Planting

None known in Lao. High potential.

TDE 65

Stem

Stem up to 50 m long and 1.2–2 cm diameter.

Sheath 2–5 cm diameter.

Sheath green. Often with stripes of white indumentum (like a Lao partridge, hence 'wai nokkhor') which is most visible on sheaths with few spines.

Spines yellow with red tips. Longest ones ≤4 cm. Very variable, some sheaths (especially low on the stem) only have short spines. All sheaths in the photo come from one plant.

Ocrea tiny, dry, with no spines or bristles

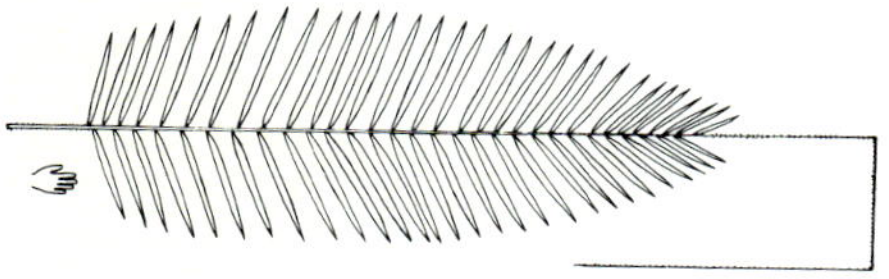

Whole **leaf** 1.5–3 m long plus a cirrus up to 1.5 m.

Petiole of upper leaves 5–20 cm long.

Largest leaflets 35–50 cm × 2–5.5 cm.

Leaflet width can vary greatly. Several prominent veins above, usually naked or with a few scattered spines. Margins bristly.

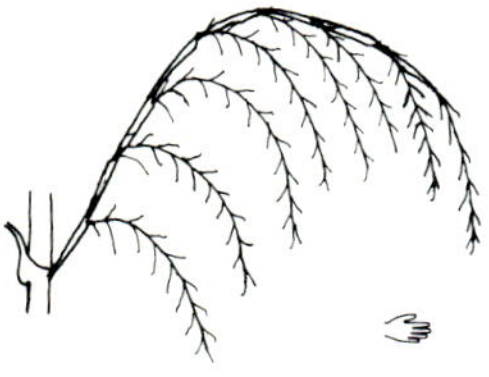

Inflorescence 0.5–3 m long with no flagellum. Erect, or arched and drooping when longer.

Primary bracts entire or with a neat dry margin. Slightly inflated, well armed.

♂ usually 3-branched.

♀ usually 2-branched.

Fruit

1.7–2.0 cm long

1.0–1.1 cm wide

Scales channelled, the margins with a narrow dark line. Perianth partly tubular.

Seed slightly ruminate.

Compare:

- *C. nambariensis* (23) and *C. obovoideus* (23a): many of leaflets grouped, larger spines always numerous, fruit ≥2.5 cm long.
- *C. platyacanthus* (24): leaflets all grouped, few or no small spines, longest spines 5–9 cm.
- *C. palustris* (22): leaflets all grouped, sheath ≤3 cm, fruit ≤1.2 cm with no grooves and the perianth not tubular.
- *C. ceratophorus* (8a): has a long, pendulous inflorescence up to 4 m long, sheath spines all long and female flower cups with stalks.

Daemonorops jenkinsiana

(boun, boun faat, wai boun, wai boun faat, wai faat, wai keedeng, wai kwa, wai seui, blong chik, re ya, kateng parua, khamay)

Climbing Clustering Cirrus Knee conspicuous

One of the commonest species. Often survives in scrub near villages, especially in the north.

Look for:

- Sheath densely covered in flat black spines. Leaf with cirrus.
- Inflorescence very distinctive, completely covered by the boat-shaped, spiny first bract.
- Leaflets regularly spaced with many black hairs on the faces.

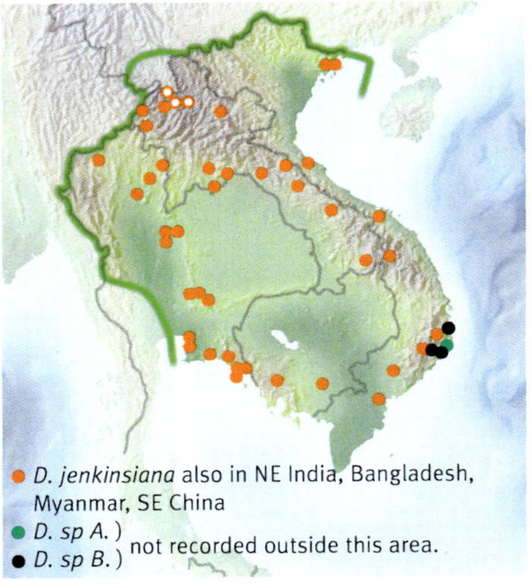

KP 315

Habitat

In Lao grows in evergreen forest and scrub at 100–1000 m, often common. In Thailand and Vietnam occurs at similar altitudes.

Fruiting and flowering months

Flowers in bud 12, 1, 3, 5 (Lao), 10, 2, 6 (Thailand), 2 (Vietnam), open 6, 10 (Thailand) Fruit unripe 12, 2, 3, 5 (Lao), 12–2 (Vietnam), 4–6, 11, 12 (Thailand), ripe 3 (Lao), 3, 5, 9, 12 (Thailand), 5 (Vietnam).

Uses

Handicrafts: suitable.
Trade: suitable.
Shoot: edible, large and delicious.
Pretty seeds exported, leaves used as thatch.

Planting

Small trial planting at Nam Xuang. High potential. Reportedly a major crop in SE China.

Stem

Stem up to 25 m long and 1.5–3 cm diameter.

Sheath 3–6 cm diameter.

Sheath green, often covered with reddish-black indumentum.

Spines many, black, a mixture of tiny needles and long flat spines.

Ocrea tiny, dry, with no spines or bristles.

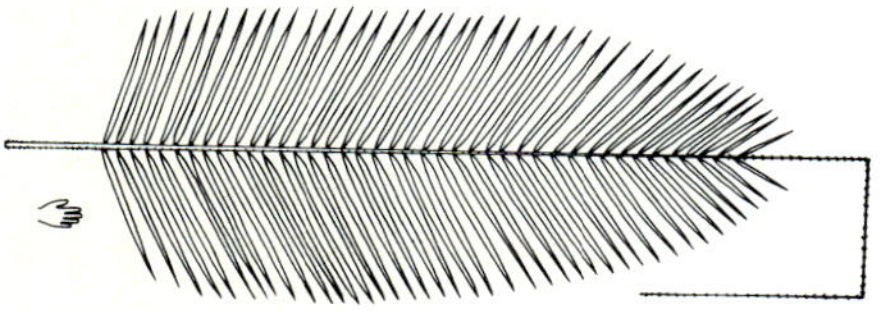

Whole **leaf** 1–2.5 m long plus a cirrus of 0.5–2 m.

Petiole of upper leaves 10–50 cm long.

Largest leaflets 30–70 cm × 1.5–3.8 cm. Middle vein above prominent. Several lines of black bristles above, even on small seedlings. Margins bristly.

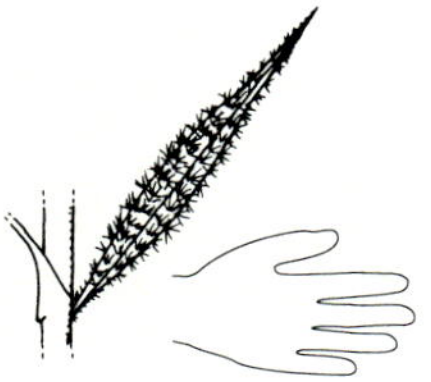

Inflorescence 0.3–0.65 m long with no flagellum. Erect. Stalk less than 5 cm long.

Primary bracts boat-shaped, nested. The outside one completely covers the inflorescence and ends in a 'beak' of variable length. Inflorescence branches of both sexes crowded, zigzag.

♂ usually 4-branched.

♀ usually 2-branched.

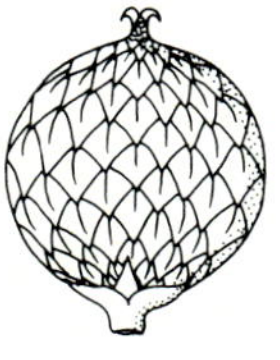

Fruit

1.9 cm long

1.9 cm wide

Scales light-coloured, channelled.

Seed ruminate.

Compare:

- Some *Calamus* species (22–25) have cirrus and knee, but the sheath spines and inflorescences are very different.
- Some *Calamus* species (18–21) have similar sheath spines but they have a flagellum not a cirrus.

• 26a *Daemonorops* sp A

Rachis and all spines with thick pale indumentum, spines at sheath mouth up to 18 cm long. Inflorescence ≥1 m long, not completely covered by one bract. Each partial inflorescence is covered by a bract which splits and then falls off before the flowers open (see picture). In *C. hypoleucus* (17) and *C. guruba* (7) the bract splits partly and does not fall off. Fruit unripe 4. Grows at 600 m.

• 26b *Daemonorops* sp B

Leaflets strongly grouped, spines lack thick indumentum. Inflorescence like *Daemonorops* sp A but shorter. Fruit ripe 4. Grows at 700 m.

27 *Korthalsia laciniosa*

(wai taleuk)

Climbing Clustering Cirrus No knee

Look for:

- Diamond-shaped leaflets, cirrus and branching stems identify the genus.
- Long, net-like ocrea, leaflets ≥25 cm long and ≥9 on each side of the leaf identify this species.
- The inflorescence has strange furry rachillae with hermaphrodite flowers.

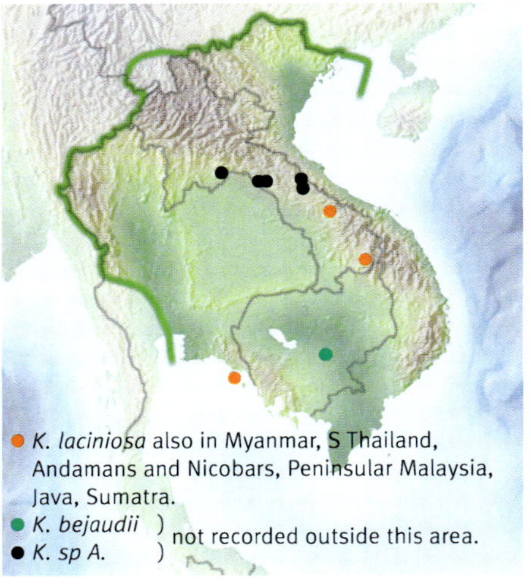

Habitat

In Lao in evergreen forest at about 200 m, in Thailand at 200 m.

Fruiting and flowering months

Flowers ?
Fruit ?

Uses

Handicrafts: unknown.
Trade: unknown
Shoot: unknown

Planting

None known in Lao.

TCW 3126 + JD 5440

Stem

Stem up to 75 m long and 1.5–2 cm diameter, branching.

Sheath up to 3.5 cm diameter.

Sheath bright green with grey-brown indumentum.

Spines sparse, the same colour as the sheath, thin, 0.6–1.2 cm.

Ocrea 10–20 cm, tattering, breaking up into a fibrous net on the side opposite the petiole. Almost unarmed.

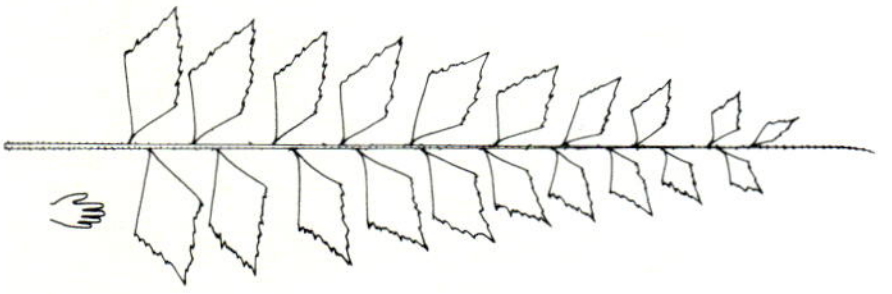

Whole **leaf** 1.5–2.5 m long plus a cirrus ≥50 cm.

Petiole 30–65 cm long (sometimes only 10 cm in S Thailand).

Largest leaflets 27–39 cm × 10–22 cm.

Broad diamond-shaped with white indumentum below. Many prominent veins which end at the margin not the tip. Leaves much smaller on sheaths with inflorescences

Inflorescence up to 0.75 m long with no flagellum. Quite erect. Stem dies after flowering.

Primary bracts entire, tightly sheathing, unarmed.

No male and female plants: each flower contains both sexes.

Usually 2–3-branched. Rachillae catkin-like, cylindrical and furry with spirally-arranged flowers.

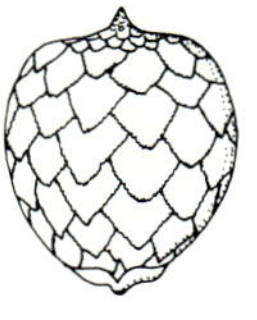

Fruit

2 cm long

2 cm wide

Scales brown when ripe, with wide pale margins.

Seed ruminate.

Compare:

- *C. bousigonii* (9a): knee and flagellum present, no cirrus, stems not branching, no ocrea, inflorescences very different.

27a *Korthalsia bejaudii*
Ocrea short (≤4 cm), entire, armed. Leaflets large (≥20 cm long), sheath about 2 cm diameter. May be a form of *K. laciniosa*.

27b *Korthalsia* sp A (wai nga, wai nyeng, wai neng, dtellhe)
Ocrea ≤9 cm (usually ≤4 cm), net-like on one side, almost unarmed. Leaflets ≤18 cm long, 4-7 per side, sheath usually ≤1.5 cm diameter. May only be a small form of *K. laciniosa*. Evergreen forest at 500–1000 m. Flowers in bud and fruit unripe 3 (Lao).

28 *Myrialepis paradoxa*

(wai namsay, wai deng, wai keekay, wai lao, reduln)

Climbing Clustering Cirrus No knee

Look for:

- Leaf sheaths and petiole near the ground have long golden spines in long combs.
- Leaflets mostly regular, never white underneath. No ocrea at any height.
- Fruits very distinctive with many tiny scales.

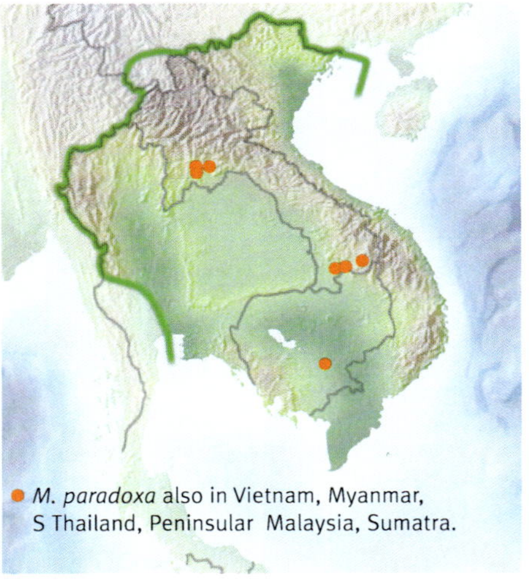

● *M. paradoxa* also in Vietnam, Myanmar, S Thailand, Peninsular Malaysia, Sumatra.

Habitat

In Lao grows in evergreen forest at 100–300 m.

Fruiting and flowering months

Flowers ?
Fruit unripe 5 (Lao).

Uses

Handicrafts: suitable (lower part of stem).
Trade: unknown.
Shoot: edible.

Planting

Small trial plantation in Nam Xuang.

KP 406 (left); no specimen (right)

Stem

Stem up to 30 m long and 1–3 cm diameter.

Sheath 2.5–7 cm diameter. Sheath green, with red-brown indumentum. Stem and sheath thin near the ground, much thicker before flowering.

Spines on lower sheaths, up to 4 cm long, golden, joined in combs of 5–10. On upper sheaths (see left) spines become shorter, more scattered.

Ocrea tiny, dry, with no spines or bristles.

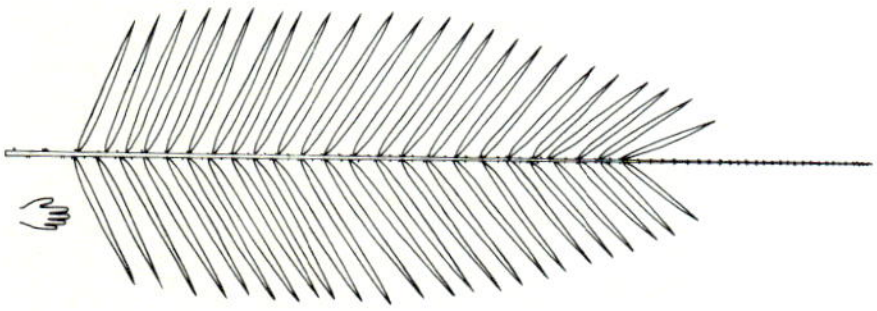

Whole **leaf** 1.5–2 m long plus a cirrus up to 1.8 m.

Petiole of upper leaves 15–30 cm long.

Largest leaflets 40–50 cm × 2.5–4 cm.

Leaflets regular or slightly grouped, with a rubbery texture. Can have a long hair-like tip. Middle veins prominent above, naked. Margins slightly bristly. Underside of leaflet with many tiny white dots (easier to see with a hand-lens).

Inflorescence 0.6–2 m long with no flagellum. Two shown in picture. Erect. Each stem flowers once and then dies.

Primary bracts entire, tightly sheathing, unarmed. Branches are nearly at right angles.

♂ usually 3-branched. When 3-branched, the rachillae are ≤2 cm long.

♀ usually 3-branched, rachillae less than 3 cm long.

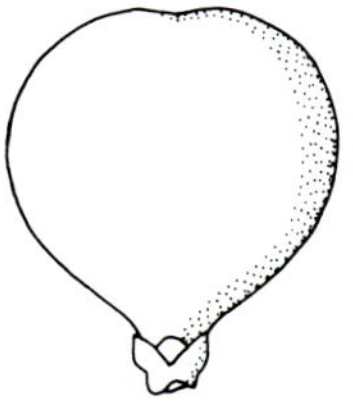

Fruit

2.5 cm long

2.5 cm wide

Scales very many, tiny, grey-green.

Seed not ruminate.

Compare:

- *P. himalayana* (31): shorter spines in longer combs, inflorescences very different.
- Other *Plectocomia* species (30–31a): leaflets white underneath when new. Inflorescences very different.
- *Plectocomiopsis geminiflorus* (29): sheath spines not in combs, ocrea ≥5 cm, papery.

29 *Plectocomiopsis geminiflora*

(wai deng, wai dengnoy, wai nou, wai nyang, wai kamlao, nyang)

Climbing Clustering Cirrus No knee

Look for:

- Leaf sheath spines ≤2 cm, scattered, not in combs.
- Ocrea ≥5 cm, papery, tattered.
- Leaflets regular, with one line of golden bristles above, never white underneath.

● *P. geminiflora* also in S Thailand, Peninsular Malaysia, Sumatra, Borneo.

Habitat

In Lao grows in evergreen forest at 200–800 m.

Fruiting and flowering months

Flowers ?

Fruit ?

Uses

Handicrafts: suitable.
Trade: unknown.
Shoot: edible.

Planting

None known in Lao.

KP 393

Stem

Stem up to 50 m long and 1–3 cm diameter.

Sheath 2–6 cm diameter.

Sheath dull green with grey or reddish-brown indumentum.

Stem and sheath thin near the ground, much thicker just before flowering.

Spines golden, scattered, 0.3–1 cm.

Ocrea 5–10 cm. Pale, papery, tatters quickly.

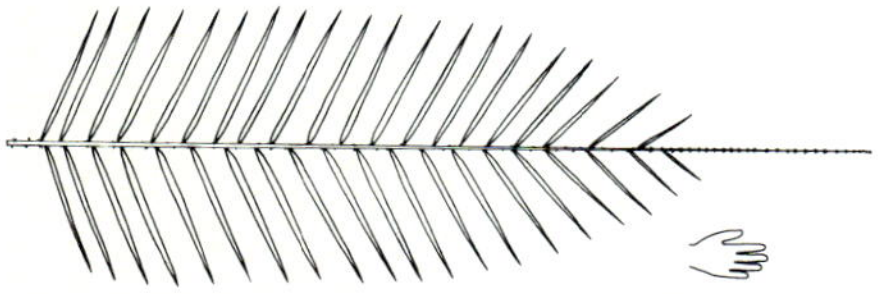

Whole **leaf** 1.1–2 m long plus a cirrus of up to 1 m.

Petiole of upper leaves 0–6 cm long.

Largest leaflets 30–40 cm × 1.5–4 cm.

Leaflets have long hair-like tips. Middle vein prominent, with a line of golden bristles. Margins bristly.

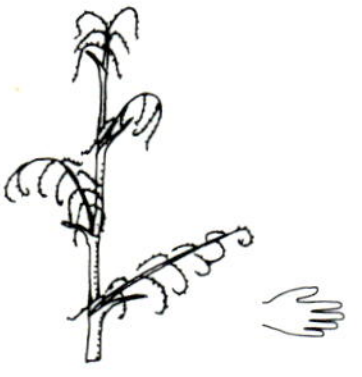

Inflorescence 0.4 m long with no flagellum. Four shown in picture. Erect. Each stem flowers once then dies.

Primary bracts entire, inflated, unarmed.

♂ usually 3-branched. Rachillae ≤0.5 cm, hardly look like rachillae at all. Branch with many rachillae shown.

♀ usually 2-branched. Rachillae very short, branch with many rachillae shown.

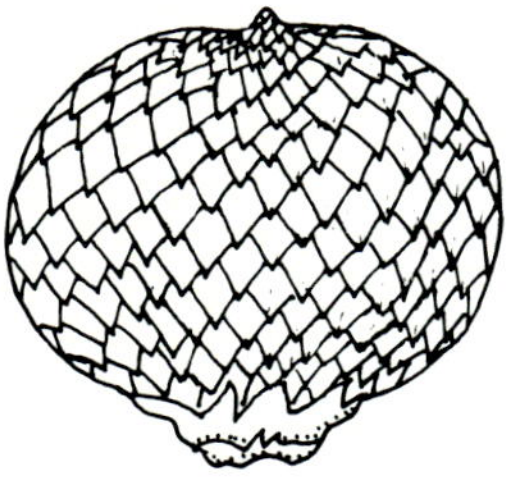

Fruit

3.0 cm long

3.0 cm wide

Top-shaped. Scales chestnut-brown with pale margins.

Seed not ruminate.

Compare:

- *Myrialepis paradoxa* (28): no ocrea, lower sheaths with spines in tall combs, no golden bristles on leaflets.
- *Plectocomia himalayana* (31): no ocrea, sheath spines in long neat combs. Inflorescences very different.
- Other *Plectocomia* species (30–31a): no ocrea, leaflets white underneath. Inflorescences very different.

30 *Plectocomia pierreana*

(wai katheui, wai lao, wai namkhao, wai tang, wai nyipoon, probably also wai hom, wai khom, wai leuk, wai loi, wai louang, wai meu, wai taven, wai thoon, wai xang, blong knair, blong roi)

Climbing Clustering Cirrus No knee

Look for:

- Leaflets white underneath (wears off with age, so check the leaflet bases on old leaves).
- Sheath spines ≤2 cm, scattered or in groups of 5–10. Petiole of upper leaves ≤3 cm long.
- Only flowering stems can be certainly identified from some other *Plectocomia* species.

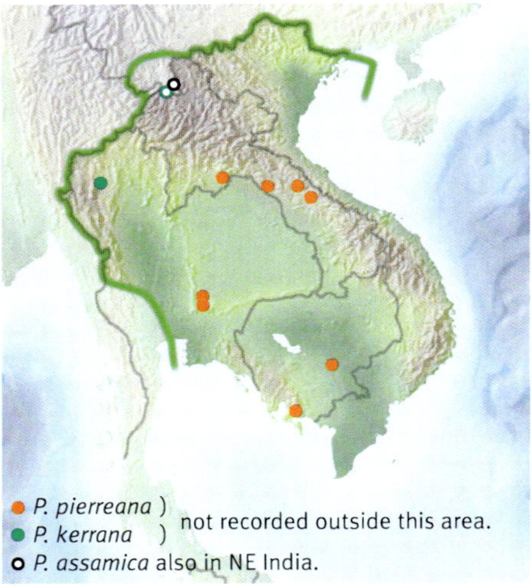

Habitat

In Lao recorded in evergreen forest at 300–550 m. Recorded at 700 m in Thailand and 900–1000 m in Cambodia. Unidentified Plectocomias, probably mainly *pierreana*, found all over Indochina up to 1000 m.

Fruiting and flowering months

Flowers open 3 (Lao), 3 (Thailand).
Fruit unripe 2, 3 (Lao), 7 (Cambodia), ripe 5 (Cambodia).

Uses

Handicrafts: suitable (lower part of stem).
Trade: not suitable.
Shoot: edible.

Planting

None known in Lao.

Stem

Stem up to 35 m long and 0.5–4 cm diameter. Stem thin near the ground, much thicker before flowering.

Sheath 1–9 cm diameter.

Sheath green, often with dark indumentum.

Spines weak, golden, needle-like. Scattered singly or in groups, usually no more than 5 together. 1–2 cm long on lower sheaths (see photo), ≤1 cm higher up.

Ocrea absent.

Whole **leaf** 2–3 m long plus a cirrus of 0.5–1 m.

Petiole of upper leaves 0–3 cm long. Up to 2.5 cm wide.

Largest leaflets 40–55 cm × 3.5–4 cm.

Newly emerged leaflets have white indumentum underneath, which gradually wears off. Leaflets with three main veins, one central and one at each margin, without bristles. Leaflet without a long hair-like tip.

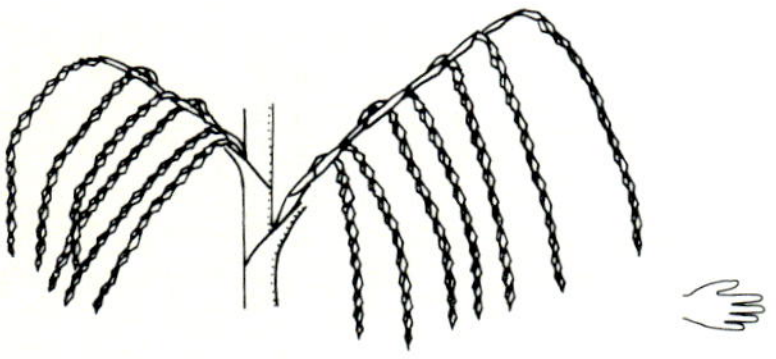

Inflorescence 0.5–1 m long with no flagellum. Two shown in picture. Branches pendulous, up to 70 cm.

Primary bracts entire, slightly inflated, unarmed. Secondary bracts overlapping, expanded to cover rachillae.

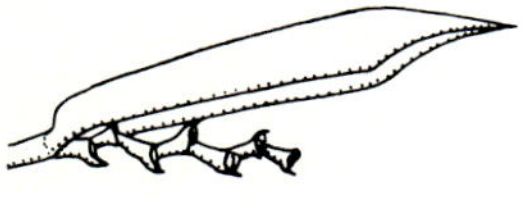

♂ 2-branched. Secondary bracts 1.9–4.0 cm long. Rachillae thin, 0.8–1.9 cm, 2.0 mm wide.

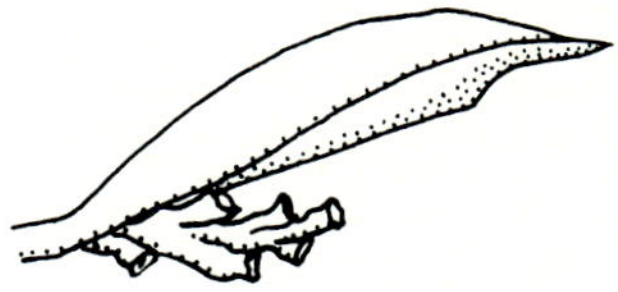

♀ 2-branched. Secondary bracts 2.5–4 cm long. Rachillae thicker than male, 1.2–2.0 cm, with red fur.

Fruit

2.0 cm long

2.0 cm wide

Scales not bristly.

Seed not ruminate.

Compare:

- *P. elongata* (31a): petiole 15–20 cm, sheath spines in long neat rows, all parts of plant bigger, fruits bristly.
- *P. himalayana* (31): leaflets not white below, tip hair-like, sheath spines in long neat rows.
- *Myrialepis paradoxa* (28) and *Plectocomiopsis geminiflorus* (29): have leaflets never white below, inflorescences very different, white dots or golden hairs on the leaflets.

• 30a *Plectocomia kerrana*

Male secondary bracts 4-5 cm long, rachillae 2.5–3.5 cm long. Female secondary bracts 4–5 cm long, rachillae ≥2.0 cm long, fruit 2.5 cm long, 2.3 cm diameter. In N Thailand at 1200 m.

• 30b *Plectocomia assamica*

Bigger (e.g. petiole over 4 cm wide). Female secondary bracts 6–7 cm long, rachillae 1.5–2.5 cm, fruit 3.5 cm diameter. Male parts unknown.

31 *Plectocomia himalayana*

(probably wai katheui)

Climbing Clustering Cirrus No knee

Look for:

- Leaflets in groups, never white below. Very long, hair-like tip.
- Sheath spines all neatly arranged in long combs (≥20 per comb). Petiole very short.
- Grows high in the mountains.

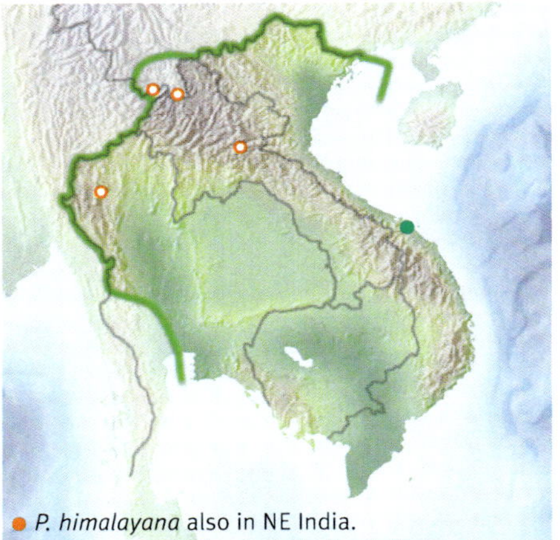

● *P. himalayana* also in NE India.
● *P. elongata*) also in S Thailand, Malaysia,) Indonesia and the Philippines.

Habitat
In S Yunnan (and probably N Thailand) grows above 1600 m.

Fruiting and flowering months
Flowers ?
Fruit ?

Uses
Handicrafts: suitable.
Trade: unknown.
Shoot: unknown.

Planting
None recorded in Lao.

NO SPECIMEN

Stem

Stem up to 10 m long and 0.5–4 cm diameter.

Sheath 1–10 cm diameter.

Sheath green.

Spines short (≤1.5 cm), yellow, all neatly arranged in long rows of 20 or more.

Ocrea absent.

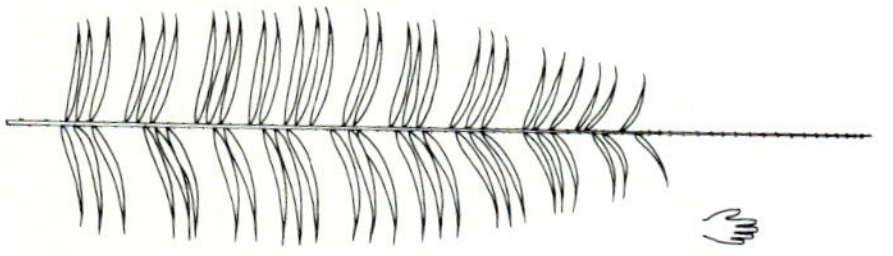

Whole **leaf** 2–3 m long plus a cirrus of 0.5–1 m.

Petiole of upper leaves 0–3 cm long. Up to 2 cm wide.

Largest leaflets 45–50 cm × 3.5–4 cm.

Leaflets never have white indumentum below and lack the white dots of *Myrialepis* (28). Five or more prominent veins on the upper surface. Tip has a long, thread-like part up to 7 cm.

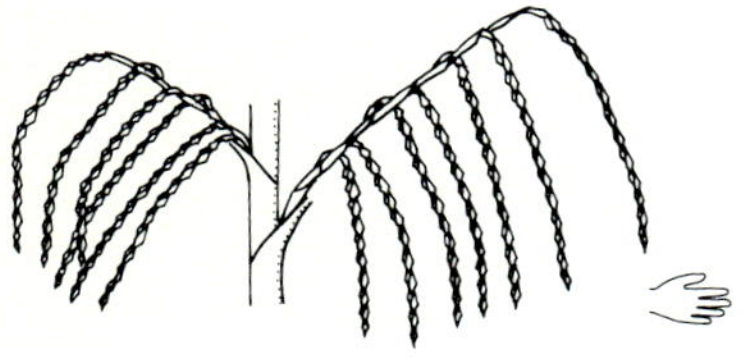

Inflorescence up to 2 m long with no flagellum. Two shown in picture. Branches pendulous, up to 80 cm.

Primary bracts entire, slightly inflated, unarmed.

Secondary bracts overlapping, expanded to cover rachillae.

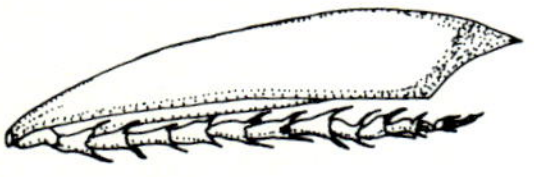

♂ 2-branched. Secondary bracts 4–5.5 cm long, rachillae 2.0–2.8 cm.

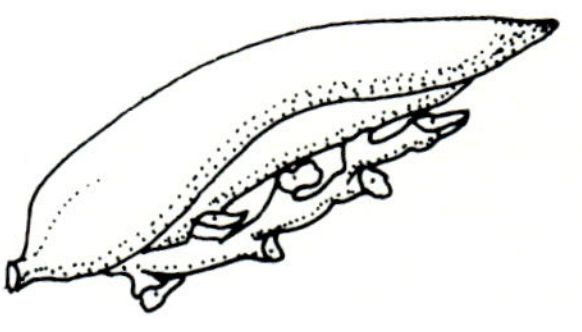

♀ 2-branched. Secondary bracts 4–5.5 cm long, rachillae 1.2–2.5 cm.

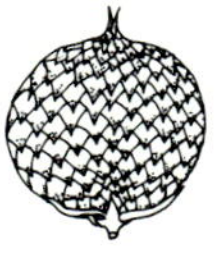

Fruit

1.5 cm long

≤1.5 cm wide

Scales not bristly.

Seed not ruminate.

Compare:

- *Plectocomia* species (30, 30a, 30b): leaflets white underneath when freshly emerged, sheath spines not in long combs, leaflets not ending in long hair-like tips
- *Plectocomiopsis geminiflorus* (29): leaflets regular, ocrea long, spines not in combs
- *Myrialepis paradoxa* (28): leaflets regular, without long tip, spines on younger sheaths and petiole in short, tall combs. Inflorescence very different.

● **31a *Plectocomia elongata***

Spines in long neat combs, leaflets white underneath, without a thread-like tip. From *P. pierreana* by larger size (e.g. petiole ≥4 cm wide, leaflets 50–65 cm), petiole 10–15 cm long, fruit bristly (see picture), 3 cm diameter. In Vietnam at 200–400 m.

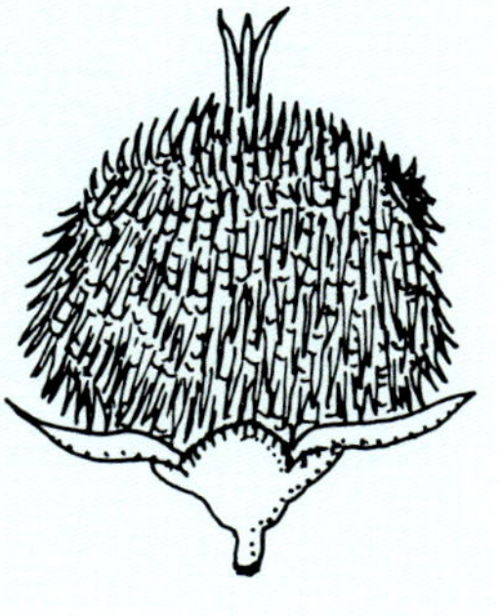

Some other plants sometimes called 'wai' in the Lao language

Salacca wallichiana Mart. ►
Sometimes called wai louang.

A big spiny palm with scaly fruits, closely related to the rattans. The stem can be over 10 cm diameter and over 1 m tall. The lower part often lies down and produces extra roots. The leaves are over 4 m long, the petiole very deeply channelled and the leaflets strongly grouped. The inflorescences are pendulous and emerge through a split in the leaf sheath. In Indochina known from the lowlands at one or two sites in Bokeo Province, NE Lao (possibly introduced), and also in SE and Central Thailand. Planted commercially in Thailand for the big, sour-sweet fruits.

OL 239

NO SPECIMEN

◄ *Pothos scandens* L.
Sometimes called wai sanoy.

A climber with no spines. The leaves look a little like the pinnate leaves of a rattan, but in fact each 'leaflet' is a separate leaf and the 'rachis' is a stem of the plant. The petiole has expanded 'wings' similar to some *Citrus* species. The inflorescence is small, purple-yellow and stands on a sharply bent stalk.

Combretum roxburghii Spreng. ►
(= *Combretum decandrum* Roxb.)
Sometimes called 'wai din'.

A woody climber with no spines, no leaf sheaths, branching, unsegmented stems, rough bark, opposite leaves, a branched head of small flowers and winged fruits. It doesn't look much like a rattan but the flexible stems are a low-quality substitute for rattan cane.

NO SPECIMEN

Parts of a rattan

The stem

Raised rings divide the stem into segments (**internodes**) like a bamboo stem but it is woody and solid, not hollow. One leaf sheath (see next section) is attached at each ring.

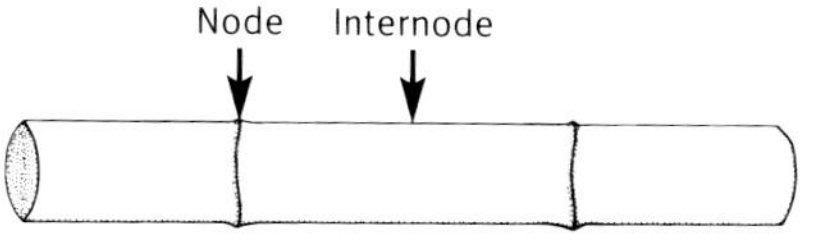

Most species have **climbing** stems so their internodes are long and flexible. A few species (e.g. *Calamus harmandii*) have short stiff internodes and an **erect** stem like a small tree. One species (*Calamus acanthophyllus*) has internodes so short that it appears **stemless**.

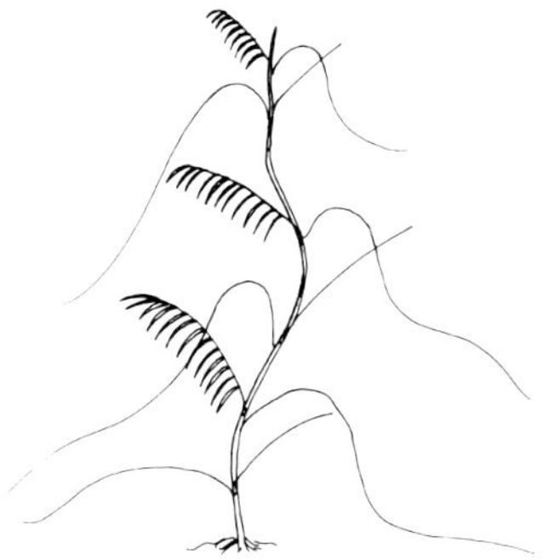

climbing stem

erect stem

stemless

Most rattan species can produce new stems from the rootstock. This is called the **clustering** habit. Some species cannot do this so an individual only ever has one stem. This is the **solitary** habit. The Lao species of *Korthalsia* can produce **branches** anywhere along the stem and other species can produce a branch occasionally if the shoot is damaged.

clustering habit

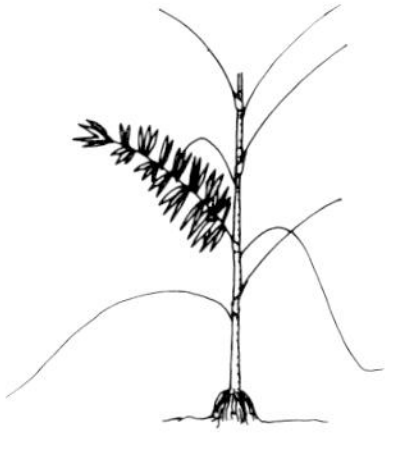

solitary habit

branching habit

The **stem length** is best measured from the ground to the base of the last fully unfurled leaf. In many climbing species this length can exceed 30 m and a few have been recorded over 100 m long. The exact length is not very helpful for identification because it often depends more on the age of the stem rather than its species. However, it seems that some species only ever produce quite short stems (e.g. *Calamus henryanus* and *Calamus hypoleucus*) which can be a clue for identification.

The new leaves and new stem segments are produced only at the tip of the stem, which is soft and fleshy. This part is often called the **shoot**. The shoots of many species are considered delicious vegetables by Lao people. Overlapping leaf sheaths conceal the shoot and upper part of the stem. Eventually the older leaves die, fall off and are not replaced, leaving the lower, woodier part of the stem bare.

Once it has become woody the **diameter** of the stem is fixed and, unlike a tree, it cannot get thicker as the plant gets older. The diameter can vary greatly between individuals or between the base and tip of a stem, but it can provide some clues for identification.

The leaf sheath

The lowest part of the leaf usually forms a tough tube, the **leaf sheath**, embracing the stem. In some species (most *Calamus* and *Daemonorops* species) there is a strong fold just below where the rachis begins. This looks like a big nose and is called the **knee**. It may be very clear or quite indistinct.

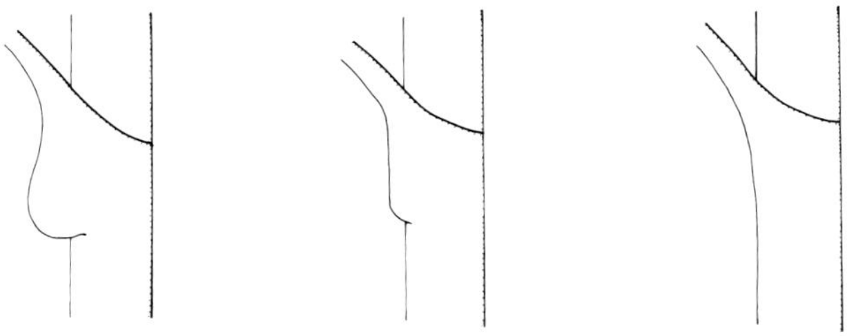

clear knee | indistinct knee | no knee

At the mouth of the sheath there is often a weak, non-woody outgrowth called an **ocrea**. In Lao species this usually takes one of the following forms:

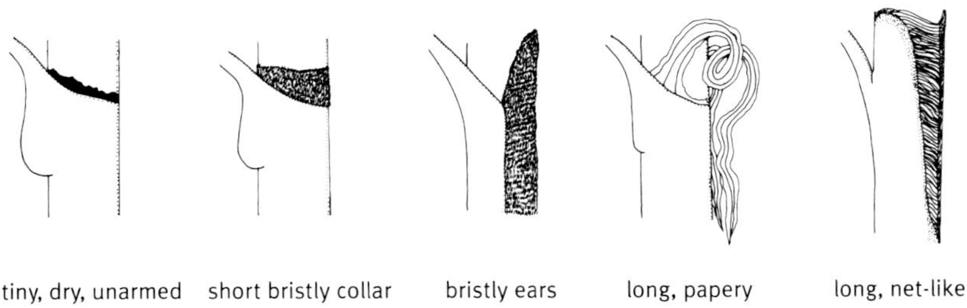

Ocrea tiny, dry, unarmed | short bristly collar | bristly ears | long, papery | long, net-like

TAKE CARE! The ocrea breaks off easily.
If you don't see an ocrea on one sheath don't assume there is never one!

Sheath spines are extremely variable. Try to look at their colour, abundance, maximum length, position, the presence of very short spines, the way they grow (singly or joined in rows) and their shape (thick, flat or needle-like). Spines on newly emerged sheaths may be bright red or yellow but on older sheaths they often become dull green. Rather than use technical words we rely mainly on pictures to show the character of the spines in each species. Note that the presence or absence of spines on the knee can vary between adjacent sheaths on a single stem in several species and is useless for identification in Indochina.

Newly emerged sheaths often have a lot of **indumentum.** This is the dusty, fluffy or waxy coating on the surface of a rattan. It is easily rubbed off with your fingers and in nature tends to be worn away as the surface gets older. On sheaths it may be white, red, brown or black and occur in patches, in neat stripes or all over. It wears off quickly except where it is protected by spines.

The leaf

The leaf is made up of a long main axis, the **rachis**, which has many **leaflets** along it. The lowest part has no leaflets and is called the **petiole**. The length of the petiole is useful for identification.

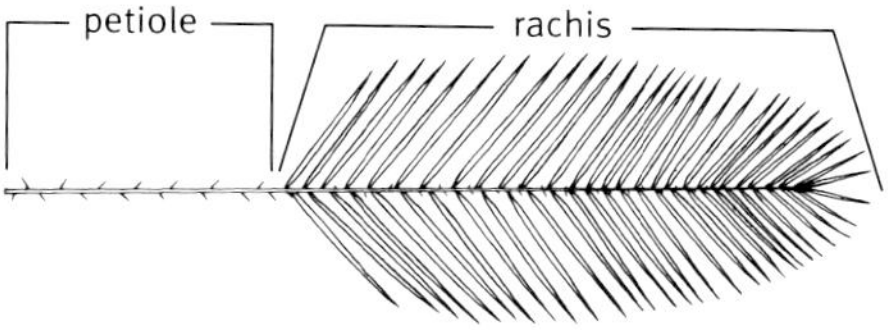

TAKE CARE! Leaves near the ground have long petioles in almost every species. **Measure the petiole on the uppermost leaves**, i.e. At least 2-3 m from the ground in slender species and more for larger ones.

The **leaflet arrangement** is a useful identification feature, although in some species more than one arrangement may be seen. Again, try to look at the upper leaves rather than those at the base of the stem. Here are some of the main types:

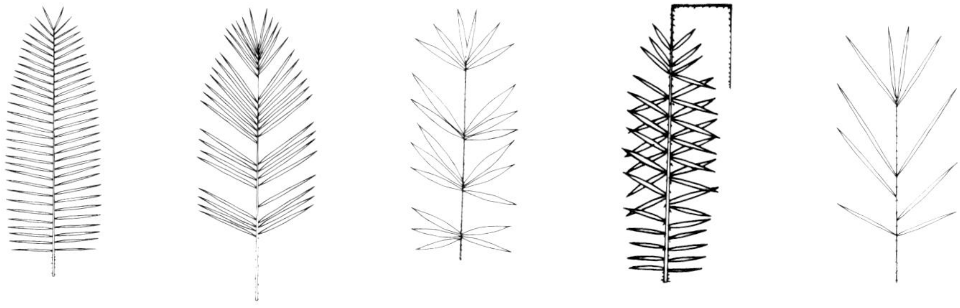

regular | interrupted pinnate | strongly grouped | some grouped, some regular | irregular

Most species have the leaflets in two neat rows so they lie **all in one plane** and the leaf looks flat overall, like a coconut leaf. A few have the leaflets pointing in all different directions (up, sideways, even down) so that they lie **in many different planes**.

The **leaflet shape** can also be distinctive, although this is even more variable. Shape is difficult to describe in words so we concentrate on using pictures in this book. One phrase we do use is **boat-shaped**. This means the surface of the leaf is very strongly curved like the hull of a boat. It is typical of *Calamus acanthospathus* and *Calamus palustris* amongst others. Most species have much flatter leaves, although all have the base strongly creased to form a roof-shape where they join the rachis.

The **leaflet size** varies greatly so to compare with the sizes in this book measure the largest leaflets you can find. When looking at the size range we give for a species remember that especially small leaves are much more likely to be found than especially large ones.

The **leaflet surface** and **leaflet margins** are often **bristly**. Look closely at where the bristles occur. The surface bristles are usually growing from the main veins – the number of main veins and the number which have bristles are worth noting for both upper and under surfaces. Most are short, black and pressed flat – take special note if they are long (≥5 mm), pale coloured or stand up.

Some species have **whitish indumentum** on the underside of the leaflets. This is a very good clue for identification.

The climbing organ

On some species the **climbing organ** grows from the tip of the leaf and is then called a **cirrus**. On other species it grows from the leaf sheath and is then called a **flagellum**. The flagellum is a modified flower head. *Calamus bimaniferus* is unique in Indochina because it has a short flagellum *and* a short cirrus.

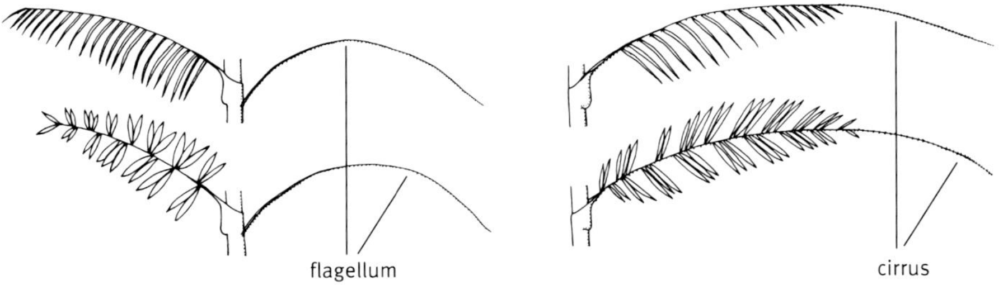

TAKE CARE! Seedlings and the lowest part of a stem do not usually have climbing organs. **Look at the highest leaves** – for bigger species that means at least 3 m off the ground.

The inflorescence

The rattan flower head or **inflorescence** is long, tough and branched. The flowers fall off quite quickly but luckily they are not required to identify the Indochinese species. The rest of the inflorescence may remain on the plant for one or two years.

Some inflorescences are quite short and **erect** with **no climbing organ**. Others are long and drooping with **pendulous branches** and a long **flagellum** reaching out beyond the flowering branches. These latter species usually also have a flagellum on non-flowering leaf sheaths.

inflorescence habits

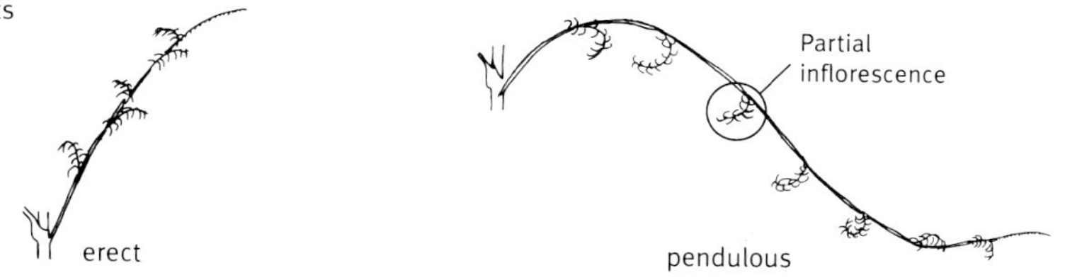

The branching pattern is quite difficult to describe. The main **axis** arises from the leaf sheath. It usually has 2-10 main side-branches. These branches may or may not branch again, and their branches may in turn branch. A few species have very simple **1-branched** inflorescences where the axis branches but these branches do not branch again (e.g. *Calamus harmandii*, female *Calamus oligostachys*). In many *Calamus* species the females are **2-branched** (meaning the axis has branches and those branches have branches) whilst the males are **3-branched** (as females, but the branches of the branches branch again). Confusingly, many species are branched more times in the part of the inflorescence near the stem than at the tip. Thus many species accounts say e.g. 1-2 or 2-3 branched, showing the range which can be found. A level-1 branch plus all its side branches is called a **partial inflorescence**.

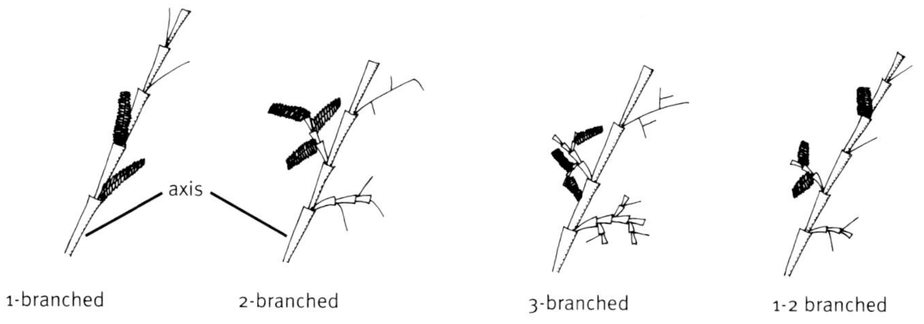

The inflorescence branches are entirely wrapped in tough tubular sheaths called **bracts**. These are often covered with spines and claws. Non-climbing species have no claws on the bracts, although some have straight spines. Some species have very **tightly sheathing bracts**, others have them quite loosely fitting or **inflated**.

Types of inflorescence bract

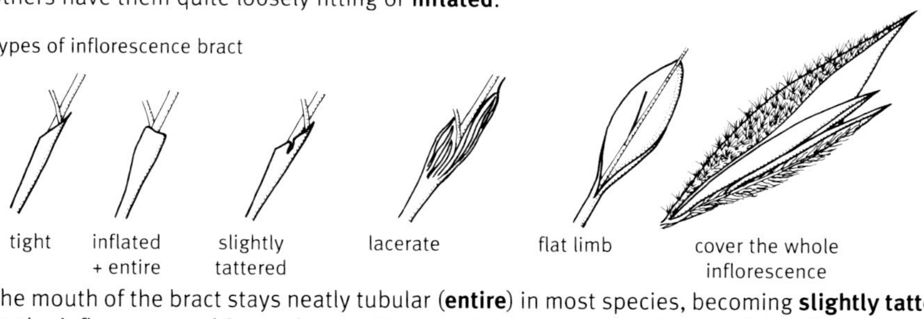

The mouth of the bract stays neatly tubular (**entire**) in most species, becoming **slightly tattered** as the inflorescence dries and ages. However, in some species the mouth breaks up into long strips and fibres (**lacerate**) and in two species (*Calamus guruba* and *Calamus hypoleucus*) the mouth opens out to form a neat, nearly **flat limb** or sheet. This can be compared with *Daemonorops* sp A and sp B where the bracts also open out but it is the **whole bract** (split right to the base) which does this, not just the mouth (see page 69). In *Daemonorops jenkinsiana* the bracts cover the whole inflorescence until flowering is over, then they split right to the base.

The axis and intermediate branches do not bear flowers – only those at the last level of branching do. A branch bearing flowers is called a **rachilla** (plural **rachillae**). The length, shape and arrangement of the rachilla are important for identification.

Rachillae (all life size)
Calamus guruba ♂ *Calamus rhabdocladus* ♂ *Calamus palustris* ♀ *Calamus flagellum* ♀

In most of the genera in this book there are separate male and female plants, but in *Korthalsia* each individual flower has male and female parts (they are hermaphrodite). In *Calamus* and *Daemonorops* the male flowers are solitary in their rows but each female flower is closely paired with an infertile male flower. Nobody knows why. This makes it easy to tell the sex of a *Calamus* or *Daemonorops* inflorescence. *C. viminalis* and *C. siamensis* often have two female flowers and one infertile male flower clustered together at each flowering site – this gives the rachilla a distinctive crowded appearance and means the fruits are often crowded in four rows instead of two.

The fruit

Ripe rattan fruits can be hard to find because they occur seasonally. However, if you do find them they are very useful for identification. The **shape** is often nearly spherical (globose) but some species are egg-shaped, top-shaped, cylindrical or ellipsoid. Most species have a short, very abrupt **beak** but in a few this is long and tapered.

Fruit shapes
globose, short beak ovoid, long beak ellipsoid top-shaped

The **size** in this book is measured as **length** (from the bottom of the fruit to the tip of the beak) and **width** at the widest point. The range given covers ripe and nearly-ripe fruit, but of course they begin life much smaller.

The **colour and pattern** of the fresh ripe fruit is important but it changes in herbarium specimens so it is poorly known for many species. The scales often have dark or pale margins which may be broad or narrow. The scales are often smooth but in several species they are **channelled** down the centre of each scale, the channels lining up to give the fruit a conspicuous ribbed appearance.

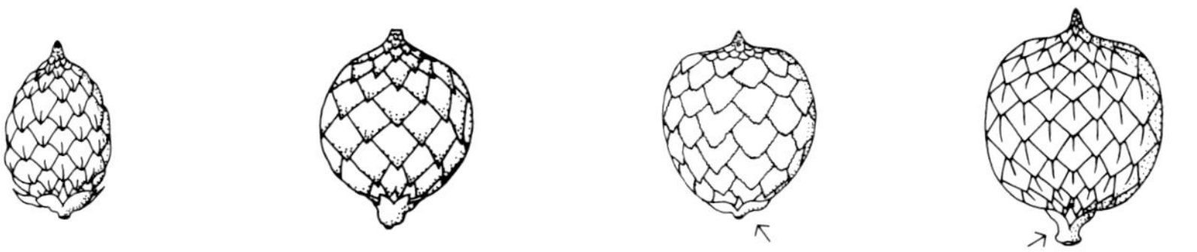

scales channelled scales not channelled perianth fully open perianth partly tubular

The leathery outer layer of the flower (the calyx) remains after pollination and forms a star-shaped cup supporting the fruit. This is the **fruiting perianth**. In some species the arms of the star only run part way to the base and the base remains **partly tubular**, lifting the fruit away from the rachilla. The perianth is not included in the length of the fruit.

Inside the seed may be plain whitish (**not ruminate**) or decorated with lots of radiating dark lines (**ruminate**).

How to collect rattans for the herbarium

It is easy to collect a bad herbarium specimen of a rattan. This is a waste of time because identification will not be possible. **Making a good specimen takes time** but so few have been collected in the past that each one you do get will form a significant addition to our knowledge. It might one day even be the basis for naming a new species. Many species (e.g. *Calamus poilanei*, *Calamus godefroyi* and *Daemonorops jenkinsiana*) were named after the first person who collected them.

Rattans are usually too large to collect whole, so you only take small parts. In particular **a complete specimen should include**:

- the leaf sheath (including any ocrea) and a 30 cm section of cane
- pieces of leaf taken from the petiole, middle and tip of the rachis. Snip off the leaflets on one side to save space but leave the bases to show their arrangement.
- the cirrus or flagellum
- the inflorescence. If it is too large cut a section of the axis which bears a major branch and all its sub-branches. Take care to include the upper part of one of the axis bracts.

Collect from the top of the plant if possible, but don't collect very immature parts (e.g. leaves which have not yet opened). Include leaves and sheaths from low down if they are markedly different. If you cut two stems from a clump look for connecting roots to check that it really is the same species.

When you cut up the rattan, each separate piece should have a numbered label attached. All the pieces you collect of one sex of one species collected in one place at one time form part of a single herbarium specimen and they should all have the same number. This number should be different from any other specimen you have collected. Rattan specimens take great delight in mixing themselves up with one another and without the label it will be impossible to be sure which piece belongs where.

Try to collect at least two complete sets of pieces. This will allow you to keep one set in Lao and send one set to a larger herbarium for critical naming if need be.

Bundle the parts in newspaper tied with string and seal them in strong plastic bags with enough **70% ethanol** to wet everything fully. On return from the field transfer them to a drying frame. Press the leaves and inflorescences but don't press the sheath and other very spiny parts – this will damage the spines.

Almost as important as the specimen are the **collection notes**. You can copy the field sheet inside the front cover of this book to use in the field. As a bare minimum note your name, the date of collection, the habitat type, the altitude and the location. Give the location in as much detail as you can – it should include country, province, district, nearest village and the name of the nearest river, mountain or other geographical feature.

Also note the appearance of the living plant. It can be difficult to look at a herbarium specimen and get an idea of what the whole plant looks like. The notes do not need to describe the pieces you have collected – people will be able to see what shape the leaflets are or how long the sheath spines are! Instead you should **write those things which cannot be seen in the specimen** – especially those things listed on the sample sheet.

If you are collecting with one or more local people who know a lot about the rattans in that area write down the names they use for the species and any information they can give you about the uses, marketability, fruiting times and other interesting features.

Managing and conserving wild rattans

Harvesting rattan can sometimes reduce populations so much that future harvests are not possible. This is particularly likely for species with solitary stems which cannot resprout after cutting. The huge solitary species *Calamus poilanei* has been virtually eliminated from very large areas of forest in Lao by commercial harvesting. Because regeneration from seed is so slow, the chance of a future harvest has been lost for a long time. The same fate has befallen the giant solitary species *C. manan* in the Thai-Malay Peninsula. One other solitary-stemmed Lao species, *C. solitarius*, is commercially important and we predict that it too will be very vulnerable to over-harvesting.

For clustering species the risk of over-harvesting is a more long-term one since the individual clumps usually survive harvesting. Prolonged heavy harvesting will prevent seed being set so new clumps will not be produced and the population will gradually age and shrink. This is probably happening over large areas of Lao forest, since harvesting is so extensive.

Heavy harvesting of young canes may also lower the economic productivity of a rattan population. There is little available information on which to base management decisions and more studies are needed, both biological and ethnobotanical. Rattans perhaps grow faster once they reach the canopy and in that case it would be more profitable to wait until they have been in the canopy for a while before cutting them. Cutting them all while they were still in the shady understorey would mean no stem ever reached its maximum growth rate.

Managing the harvest of wild rattans will be difficult for many reasons, in particular the lack of legal tenure over forest resources. This means people from one village trying to manage a stand of rattans for the future cannot exclude people from elsewhere who wish to cut them all immediately. Other difficulties include the lack of cultural precedent for managing forest resources, the very large areas of forest involved, the low staffing levels of district forestry offices or protected areas and the poverty of many rural people which forces them to over-harvest in the short term just so that they can survive. The national programmes of land allocation, management around protected areas and improving rural education may all help but it is still going to be very hard to avoid rattans being completely over-harvested in most areas.

The Lao government has declared a large system of protected areas. Managing them is very difficult, but if they are well protected they could potentially ensure the long term survival of most Lao rattan species and most other plant species in the country. Most of the forest outside them will probably be cleared or have their rattans over-harvested in the future, so protected areas will be the main sources of new species and seeds for cultivation.

Ten Indochinese species are listed as Threatened in the 1997 Plants Red Data Book. Two, *Calamus harmandii* and *Calamus poilanei*, are known from Lao whilst most of the rest are only known from Vietnam. Now there is better information from Lao this needs to be reviewed. As noted above, *Calamus solitarius* should perhaps also be considered for this list. Lao also has a special responsibility for *Calamus laoensis* and *Calamus bimaniferus* which are known only from Lao and for *Calamus oligostachys* and *Calamus hypoleucus*, each of which is known from only two sites, one of them in Lao.

Planting

Forest clearance, population growth and heavy harvesting are reducing the supply from the wild at a time of increasing demand. This is making plantations more economically viable. Commercial planting has already begun for shoot production, perhaps because convenient wild sources of this perishable vegetable have already been exhausted within reach of the main Lao towns. There is one *C. viminalis* plantation producing shoots in Bokeo Province and one village in Salavan Province where *Daemonorops jenkinsiana* is planted. *C. tenuis* has been grown in the Vientiane area since 1994 and is so successful that by late 1998 there were dozens of growers with over 70 ha of plantations in several provinces. The spread has continued since then. These plantations are grown in open fields with stocking of about 1600 plants/ha and can be very productive on good sites – reportedly up to 1 shoot per plant per month can be harvested (see the next section for more details). The inspiration to plant *C. tenuis* in Lao came from north-eastern Thailand where shoot plantations were first tried in 1991 and now cover several thousand hectares. In Thailand *C. viminalis* is the preferred species (although the plantation owners call it *C. siamensis*) with some true *C. siamensis* and *C. tenuis* also planted.

Commercial plantations for stem production do not exist in Lao. They cannot yet compete with the wild stocks which remain in remote areas. Also, cane plantations give a slow return. They can take 6-20 years to yield a harvest, depending on the species, compared with 1-2 years for shoot production. Nonetheless the Lao government can foresee cane shortages in the future and is supporting small plantation and nursery trials at the Forestry Research Centre's field station at Nam Xuang, near Vientiane (see next section).

In Ban Phon Pao, Tulakhom District, Vientiane Province there is a successful 3 ha *C. viminalis* rattan garden. It is not a true plantation since most of the plants are haphazardly self-seeded. The area is managed to encourage the rattans and a steady supply of canes, shoots and fruits is harvested for sale. The garden also produces grazing for livestock, firewood, mushrooms, fruit, and medicinal plants and it may be a useful model for integrating rattan into village gardens and agroforestry systems elsewhere in Lao.

In South Thailand there are small trial plantations of at least four species. In South Yunnan trials and government subsidised commercial plantations concentrate on *C. yunnanensis* (= *C. acanthospathus*) and *C. nambariensis*. In Vietnam there are reportedly thousands of hectares of rattans planted to produce stems for home use and sale, often as small areas along fences or in home gardens. The main species planted is reportedly *C. tetradactylus* (which has been planted for over 100 years) with smaller amounts of species given the names '*C. armarus*' (probably the same as *C. tenuis*), *C. platyacanthus*, *C. rudentum*, *C. poilanei*, *C. palustris*, *C. dioicus* (possibly *C. salicifolius* or true *C. dioicus*) and *C. tonkinensis* (which is the same as *C. walkeri*).

Cultivation techniques

What to plant and where

Rattans can be planted to produce stems or shoots. For shoot production *Calamus viminalis*, *C. siamensis* and *C. tenuis* grow well in central Indochina. Others may be suitable for different areas. For example, *Daemonorops jenkinsiana* may be suitable in hilly areas below 1000 m, including places at risk of occasional fires, and *C. nambariensis* may be good in very high altitude areas of northern Indochina.

No stem plantations have been tried in Lao so it is difficult to recommend which species to plant where, but some promising species are listed in the table below. *C. tetradactylus* and *C. platyacanthus* are especially recommended by Vietnamese scientists for planting to produce

Typical diameter	High quality	Moderate quality or quality uncertain
Large (≥2 cm)	*Calamus poilanei, C. rudentum*	*C. laoensis*
Medium (1-2 cm)	*C. nambariensis* *C. platyacanthus* *C. wailong*	*C. viminalis, C. siamensis, C. tenuis* *C. godefroyi, Daemonorops jenkinsiana,* *Myrialepis paradoxa,* *Plectocomiopsis geminiflora*
Small (≤1 cm)	*C. acanthospathus, C. gracilis,* *C. solitarius, C. tetradactylus,* *C. palustris*	*C. guruba, C. henryanus,* *C. oligostachys, C. bimaniferus*

stems. We have not seen their specimens of these so we are not certain they are the same as the species with these names in this book, but the descriptions they give sound very similar.

Plantations producing shoots do very well in open fields with no shade, at least for the three lowland species mentioned above. Plantations for stems require supports for climbing, so they are usually put in tree plantations, orchards, logged or regenerating forest. Studies in Malaysia and Indonesia show that stems grow fastest with quite a lot of light; about 50% canopy cover seems ideal for many species. To achieve this select an area with few large trees, otherwise you may need to kill some of the canopy trees in the area you are planting.

Almost nothing is known of the soil preferences of Lao rattans, although *C. tenuis* seems to grow best in moist, alluvial soils which flood every year. Preferred altitudes can be estimated from the altitudes where the species occurs naturally. For example, wild *C. tenuis* has only been found below 300 m and wild *C. viminalis* has only been found below 600 m, so planting these species at higher altitudes will probably be difficult. Most species will probably grow best in provinces where they grow naturally, and from locally collected seed which will be fresh and adapted to local conditions.

Nursery techniques

Collecting seedlings from the wild gives very poor results for rattans. In North Vietnam people reportedly propagate one species ('*C. armarus*', probably the same as *C. tenuis*) by digging up parts of the mature rootstock and this is also done sometimes in Thailand when seed is not available, but this is unusual and only works on a small scale. It is usually best to grow from seed. In Lao nursery techniques are best developed for *C. tenuis* but are still being improved by experimentation. The following advice refers mainly to *C. tenuis*.

Firstly, it is important not to confuse this species with *C. viminalis* which grows in similar areas and is often also called *nyair* or *wai nyair* in Lao (see pages 24–25 and 28–29). In Lao *C. tenuis* begins to flower during September to November and the fruits begin to ripen during March through to May. Ripe fruits have a skin which is whitish and a flesh which is reddish-brown with a sweet-sour taste. They are then ready for propagation. Collecting later in the season ensures the fruits are ripe and also means weather conditions are cooler and wetter, which is good for germination.

When collecting the fruits you should climb a nearby tree and cut the fruiting branch required. You should tie a string to the branch before cutting so that it does not fall and knock off the fruits. Once cut, the fruits should be removed from the branch. If you need to transport them, put them in a bag or sack which lets the air circulate, never a plastic bag. If you cannot plant them immediately they should be stored in a place which remains humid. Do not store them for more than a few days or germination will be poor.

Two alternative preplanting treatments can improve germination percentage and speed. For large quantities it is easiest to soak them in cold water for 7-10 days. Change the water every day to prevent mould. The skins and soft flesh should be removed, either before or after soaking. The second method is only practical for smaller quantities. After removing the skin and flesh, use a sharp pointed knife to remove a small cap which covers the embryo (also called the hilar cover). Take care not to cut the embryo.

Seedbeds are usually used for germination. However, we have found that it is easier for *C. tenuis* to mix the treated seeds with a deep bed of coconut fibres in a plastic bowl and keep them moist and covered in a cool place. Once they have germinated you can easily pick them out and plant them on. If you want to use a seedbed it should be aligned north-south and roofed with a net or coconut leaves to give about 50% shade. The soil should be turned over and carefully weeded a week before sowing. Raise the level of the bed 15-20 cm with a mix of 1 part sand to 3 parts soil.

Water the bed thoroughly, scatter the pre-treated seeds across the surface and then cover with a 1.5-3 cm layer of rice husks, sand or soil. This keeps water near the seeds and also stops them being washed away during watering. Water every day.

The seeds will begin germinating after about 15 days (untreated seeds would take 3-5 months) and continue to emerge little by little until 90 days, by which time almost all have sprouted. The first leaf appears when the seedling is one month old, at which time it will be 3 cm tall with a small root. This is the ideal time to transplant to polybags.

The polybag nursery should also have a net providing 50% shade. After one month you should remove the net so the seedlings can develop sunlight tolerance. Water the seedlings thoroughly before transplanting, transplant on a cool, cloudy day and separate the seedlings carefully with a stick to avoid damaging them.

You should make a hole about 10-15 cm deep in the soil of the bag the seedling is planted into. Plant the seedling upright then press the soil down firmly all around it. A usual size of bag is 3/5 × 7 to 4/6 × 10, and the medium one part sand to three parts soil.

Take good care of the seedlings, watering them daily if it is not raining. So far no serious losses due to disease have been reported from Lao, but to be safe remove diseased or dead seedlings quickly. Weed regularly to prevent weeds overtopping the seedlings. Once the seedlings have reached 45 days old we find it useful to add N:P:K 15:15:15 fertiliser, (one soupspoon per 20 litres of water every 15 days), and increase this to two spoons when the seedlings are about four months old. Stop feeding two months before transplanting.

Plantation techniques

When the seedlings are 9-12 months old they will be strong and able to tolerate growing outside, so this is a suitable time to plant them out. At this time the rainy season should have begun so the soil will be good and moist. Suitable soil will be quite rich, a mix of sand and loam. If possible it should be close to water. This will be good because *Calamus tenuis* likes to grow in moist areas and it makes watering easier. Clear the area, plough a week before planting and use a spacing of about 0.5–1 × 2.5 m.

We suggest adding some N:P:K 15:15:15 fertiliser in the bottom of each hole before planting, about 3 soupspoons per hole. Take care not to break the ball of earth when removing the polybag because this will kill the seedling. Put the seedling in the middle of the hole, add more earth and tread it firmly.

After planting, water the plantation morning and evening for 2-3 days so the seedlings take and grow well. After that you can reduce it to once per week. After 2 months we find it useful to put N:P:K 46-0-0 or 18-46-0 fertiliser on at about 2 spoonfuls per plant, about 25 cm from the plant, or manure (about half the volume of the planting hole per plant). After fertilising you should water or the plant will die. When the plants have been out one year you should start to add fertiliser at the rate 1-2 g / plant / 2 months. Small grasshoppers may visit the plants, but this is unlikely to be a serious problem.

Shoots can be cut from the age of approximately eight months after planting (depending on the site and weather), by which time the clumps should have 4-5 shoots each. The first cut should be the longest stem; two weeks later you can probably cut again. On average a single clump can be harvested up to once per month. If you irrigate production is year-round, if not production is limited to the seven months of the rainy season.

You should cut shoots which are 1-1.5 m tall. Pick shoots which have the last exposed sheath about 10-15 cm long because this is when the edible part of the shoot is longest. Cut about 25 cm above ground level, because this seems to improve the growth of new shoots. Shoots stay fresh enough to sell for several days. Shoots for sun-drying are steamed first, which ensures quicker drying and prevents them going black.

The above method is probably adequate for other species but for the best results it may be necessary to change some details. Work on this subject is not complete but initial nursery trials with some other Lao rattan species have given the following results:

C. viminalis and *C. siamensis* – need to be pre-treated for good germination. Soaking for a week is effective. Removing the embryo cover works well in Thai studies.

C. solitarius – soaking the seeds for a week greatly improves germination.

C. poilanei and *Daemonorops jenkinsiana* – soaking improves the germination percentage but it remains slow (it can take 4-6 months).

*Myrialepis paradox*a – 90% germination was achieved with no pre-sowing treatment, so treatment is probably unnecessary.

In South Yunnan at high altitudes *C. yunnanensis* (= *C. acanthospathus*) and *C. nambariensis* germinate well in nurseries with no pre-sowing treatment.

Detailed information for the species planted in Vietnam can be found in the book by Dung and Guang (1996) and for Thailand in the book by Kundilok and others (1997). These are listed in the section on Further Reading. The methods are generally similar to those described above, although the details vary for each species. *C. platyacanthus* is quite difficult to germinate so they recommend soaking the seeds in dilute sulphuric acid or hot water for a few minutes to encourage them. Both books state that seeds of some species can be stored for up to three months if kept in a fridge or in moist sand. In Vietnam they recommend planting for stem production in various places: home gardens, living fences, logged forest, stream sides, forest edges and scrub growing up after shifting cultivation. In scrub you need to plant some tall trees to provide shade and support in the future. In logged forest they suggest that you plant in cleared strips with 2-3 seedlings together in each spot, and make sure there is enough light by continually trimming trees and bushes.

Checklist of Indochinese rattans and alternative names

After each species name we list the person who first used it publicly. Becc. = Beccari.

1 **Calamus acanthophyllus** Becc.
2 **C. harmandii** Pierre ex Becc.
2a **C. oxycarpus** Becc.
2b **C. sp A**
3 **C. erectus** var. **erectus** Roxburgh
3a **C. dongnaiensis** Becc.
3b **C. sp B**
4 **C. viminalis** Willdenow
4a **C. thysanolepis**
var. **thysanolepis** Hance
5 **C. siamensis** Becc.
5a **C. walkeri** Becc.
6 **C. tenuis** Roxburgh
6a **C. godefroyi** Becc.
7 **C. guruba** Buchanan-Hamilton
8 **C. poilanei** Conrard
8a **C. ceratophorus** Conrard
9 **C. acanthospathus** Griffith
9a **C. bousigonii** Becc. ssp. **bousigonii**
10 **C. gracilis** Roxburgh
11 **C. henryanus** Becc.
12 **C. solitarius** T. Evans *et al.*
13 **C. oligostachys** T. Evans *et al.*
14 **C. tetradactylus** Hance
15 **C. kingianus** Becc.
15a **C. dioicus** Loureiro
16 **C. bimaniferus** T. Evans *et al.*
17 **C. hypoleucus** (Kurz) Kurz
17a **C. salicifolius** Becc.
18 **C. laoensis** T. Evans *et al.*
18a **C. longisetus** Griffith
19 **C. flagellum** Griffith var. **flagellum** and
var. **furvifurfuraceus** S.J. Pei & S.Y. Chen
20 **C. rudentum** Loureiro
21 **C. rhabdocladus** Burret
22 **C. palustris** Griffith
var. **cochinchinensis** Becc.
23 **C. nambariensis** Becc.
23a **C. obovoideus** S.J. Pei & S.Y. Chen
24 **C. platyacanthus** Warburgh ex Becc.
25 **C. wailong** S.J. Pei & S.Y. Chen
26 **D. jenkinsianus** Martius var. **jenkinsianus**
26a **D. sp A**
26b **D. sp B**
27 **Korthalsia laciniosa** (Griffith) Martius
27a **K. bejaudii** Gagnepain ex Humbert
27b **K. sp A**
28 **Myrialepis paradoxa** (Kurz) J. Dransfield
29 **Plectocomiopsis geminiflora** (Griffith) Becc.
30 **Plectocomia pierreana** Becc.
30a **P. kerrana** Becc.
30b **P. assamica** Griffith
31 **P. himalayana** Griffith
31a **P. elongata var. ?** Martius ex Blume.

Equivalent, alternative names which have been used in Indochina:

2 *Zalacella harmandii* (Pierre ex Becc.) Becc., **4** var. *fasciculatus* Becc. and subvars, **5** var. *malaianus* Furtado, **5a** *C. tonkinensis* Becc., **7** *C. multirameus* Ridley, **9** *C. feanus* Becc. and *C. yunnanensis* (and all its vars) S.J. Pei & S.Y. Chen, **11** *C. balansaeanus* Becc., **13** *C. pauciflorus* T. Evans *et al. non* Ridley, **14** var. *bonianus* (Becc.) Conrard, *C. cambojensis* Becc. and *C. bonianus* Becc., **17a** var. *leiophyllus* Becc., **19** var. *karinensis* Becc. and *C. karinensis* (Becc.) S.J. Pei & S.Y. Chen, **21** *C. pseudoscutellaris* Conrard and var. *cylindrocarpus* Conrard, **22** *C. loiensis* Hodel and *C. kerrianus* Becc., **23** the five vars named by S.J. Pei & S.Y. Chen, **24** var. *mediostachys* S.J. Pei & S.Y. Chen and *C. giganteus* var. *robustus* S.J. Pei & S.Y. Chen, **26** *D. schmidtianus* Becc., *D. pierreanus* Becc. and *D. margaritae* (Hance) Becc., **26a** *D. longispathus* Conrard non Becc. var. *mollispina* Conrard, **26b** *D. geniculatus* Conrard non (Griffith) Martius, **27** *K. grandis* Griffith, **28** *M. floribunda* (Becc.) Gagnepain and *Bejaudia cambodiensis* Gagnepain, **30** *P. cambodiana* Gagnepain and *P. barthana* Hodel

Acknowledgements

This book was researched and produced with funding from the Darwin Initiative for the Survival of Species (operated by the Department of the Environment of the UK government) through the project 'Diversity and Sustainable Use of Rattans in Lao PDR'. This is a collaborative project between the Oxford Forestry Institute, Department of Plant Sciences, Oxford University, UK, the Royal Botanic Gardens, Kew, UK and the Forestry Research Centre (FRC), National Agriculture and Forestry Research Institute (NAFRI), Vientiane, Lao PDR.

We gratefully acknowledge the approval and support of the many members of the Lao government, in particular Dr Siene Sapangthong, Ministry of Agriculture and Forestry (MAF), Khampiou Vissapra, Deputy Director-General of the Cabinet Office, MAF, Khampheuane Kingsada, Director-General of the Department of Forestry, Dr Ty Phommasack, Director-General, NAFRI, Khampay Manivong, Director, FRC, Vongxay Manivong, Sounthone Ketphanh and the staff of the Foreign Investment Management Committee, the Ministry of Foreign Affairs, NAFRI, FRC, the IUCN Non-Timber Forest Products Project, the CARE Non-Timber Forest Products Information Centre and the many Provincial and District Forestry Offices we worked with. We also extend a general thank you to the huge number of Lao people who provided accommodation, assistance and information during the fieldwork, in particular the people of Ban Naphong and the family of Mae Gokham in Ban Phon Pao.

The foundations for this book were laid during 1992-1997 by a team led by Sounthone Ketphanh during the Lao Bamboo and Rattan Research Project, which was funded by the Canadian government through the International Development Research Centre.

The project has been supervised by Dr Nick Brown and Dr John Dransfield. We have also received a great deal of assistance from many other people, including Dr Bill Baker, Peter Boyce, Joost Foppes, Khamlek, Yanyong Kongkhram, Vichit Lamxay, Dr James Maxwell, Dr Rungnapar Pattanavibool, Pheng Phengsitham, Dr Sakonsak Ramyarangsi, Prof. Chen Sanyang, Lucy Smith, Dr Isara Vongkaluang, Dr Tim Whitmore, Rosemary Wise and the staff of the following herbaria: Royal Botanic Gardens, Kew, Arnold Arboretum, Bailey Hortorium, Kasetsart University, Bangkok, Royal Forest Department, Bangkok, Guangxi Institute of Botany, South China Botanical Institute, Institutes of Ethnobotany and Botany, Kunming, Leningrad, Musée de Phanerogamie, Paris, Institute of Botany, Beijing and Stockholm.

We especially thank John Rowntree for the extraordinary welcome he gave to Lao members of the team during their stays in the UK.

All the authors wish to thank their families for their support and in particular for tolerating their repeated long absences during fieldwork.

Further reading

Taxonomy

Alam, M.K. (1990) *Rattans of Bangladesh*. Bangladesh Forest Research Institute, Chittagong, Bangladesh.

Basu, S.K. (1992) *Rattans (Canes) in India. A monographic revision*. Rattan Information Centre, Kepong, Kuala Lumpur.

Beccari, O. (1908) Asiatic palms – Lepidocaryoideae. Part I. The species of *Calamus*. *Annals of the Royal Botanic Garden, Calcutta* 11(1): 1–518.

Beccari, O. (1911) Asiatic palms – Lepidocaryoideae. Part II. The species of *Daemonorops*. *Annals of the Royal Botanic Garden, Calcutta* 12(2): 1–237.

Beccari, O. (1913) Asiatic palms – Lepidocaryoideae. Supplement to Part I. The species of *Calamus*. *Annals of the Royal Botanic Garden, Calcutta* 11(1) (App.): 1–142.

Dransfield, J. (1979) *A manual of the rattans of the Malay Peninsula*. Malaysian Forest Records No. 29, Forest Department, Kuala Lumpur.

Dransfield, J. (1981) A synopsis of the genus *Korthalsia* (*Palmae: Lepidocaryoideae*). *Kew Bulletin* 36: 163–194.

Dransfield, J. (1982) A reassessment of the genera *Plectocomiopsis*, *Myrialepis* and *Bejaudia* (Palmae: Lepidocaryoideae). *Kew Bulletin* 37: 237–254.

Dransfield, J. (in press) Two new species of *Daemonorops* (*Arecaceae: Calamoideae*) from Vietnam. *Kew Bulletin*.

Evans, T., and Anh, T.P. (in press) A new species of *Calamus* (*Arecaceae: Calamoideae*) from Vietnam. *Kew Bulletin*.

Evans, T., Sengdala, K., Viengkham, O.V, Thammavong, B. and Dransfield, J. (in press) *Calamus pauciflorus* T. Evans *et al. non* Ridl. renamed *C. oligostachys* T. Evans *et al. Kew Bulletin*.

Evans, T., Sengdala, K., Viengkham, O., Thammavong, B. and Dransfield, J. (2000) Four new species of *Calamus* (*Arecaceae*: *Calamoideae*) from Laos and Thailand. *Kew Bulletin* 55: 929–940.

Evans, T., Sengdala, S., Viengkham, O., Thammavong, B. and Dransfield, J. (in prep.) A synopsis of the rattan flora of Laos and neighbouring areas of Indochina.

Gagnepain, F. and Conrard, J. (1937) Palmiers. pp 946–1056 *in* Lecomte, H. (ed.) *Flore Générale de l'Indochine*. Vol. 6 (parts 7 & 8). Massie and Co, Paris. [In French]

Greuter, W. (editor) (2000) *International code of botanical nomenclature*. Koeltz Scientific Books, Konigstein, Germany.

Hodel, D. (ed.) (1998) *The palms and cycads of Thailand*. Kampon Tansacha, Nong Nooch Garden, Thailand and Allen Press, Kansas, USA.

Madulid, D. (1981) A monograph of the genus *Plectocomia*. *Kalikasan* 10(1): 1–94.

Pei, S.J. and Chen, S.Y. (1989) New materials of Palmae from China. *Acta Phytotaxonomica Sinica* 27(2): 132–146. [In Chinese]

Pham-hoang Ho (1993) *Cayco Vietnam* [=Flora of Vietnam] III(2): 496–527. [In Vietnamese]

Other subjects

Dransfield, J. and Manokaran, N. (eds.) (1994) *Rattans*. PROSEA Foundation, Bogor, Indonesia.

Evans, T.D. (2000) The rediscovery of *Calamus harmandii*, a rattan endemic to Southern Laos. *Palms* 44(1): 29–33.

Evans, T.D. and Sengdala, K. (1999) From non-timber forest product to cash crop - the recent spread of rattan cultivation for edible shoot production in the Lao PDR. *Lao Journal of Agriculture and Forestry* 1999(2): 40–47.

Evans, T.D. and Sengdala, K. (2001) The Indochinese rattan *Calamus acanthophyllus* – a fire-loving palm. *Palms* 45: 25–28.

Evans, T.D. and Viengkham, O.V. (in press) Inventory time-cost and statistical power: a case study of a Lao rattan. *Forest Ecology and Management*

Johnson, D. (1997) *Non-wood forest products: tropical palms.* RAP Publication 1997/10 Non-wood forest products, FAO, Bangkok.

Kundilok, S., Semsanthat, N. , Atthiratbanya, W. and Nitiwansay, W. (1997) [*Rattan.*] Bamboo and Rattan Research Group, Silvicultural Garden, Technical Forestry Division, Department of Forestry, Bangkok. [In Thai]

Rao, A.N. and Rao, V.R. (1997) *Rattan – taxonomy, ecology, silviculture, conservation, genetic improvement and biotechnology.* Proceedings of Training Courses-cum-Workshops, 14-26 April 1996, Sarawak, Sabah. IPGRI-APO, Serdang, Malaysia.

Subansanee, W. (1995) [*Forest products of Thailand.*] Publication T451, Technical Forestry Department, Royal Forestry Department, Bangkok. [In Thai]

Wan Razali, W.M., Dransfield, J. and Manokaran, N. (1992) A *guide to the cultivation of rattan.* Forest Research Institute Malaysia, Kepong, Kuala Lumpur.

There are also useful articles in the journal *Palms* (formerly *Principes*) and the *Newsletter* of INBAR (the International Network for Bamboo and Rattans).

Leaflets on rattan cultivation are available from the Forest Research Centre, Vientiane.

Rattans of Lao PDR Field specimen record sheet

Genus/species:

Collector name | Additional collectors

Collector prefix and number | Date of collection:

Country LAO PDR | Province
District | Village
Place | Altitude m
Latitude (North) | Longitude (East)

Habitat notes:

Wild/cultivated

Habit: climber/immature/not climber | Male/Female

Number of stems: | Longest stem: m

Sheath diameter cm

Sheath colour Green/Yellow/Green+yellow mixed/Dark/Light/Other notes

Stem diameter (no sheath) cm

Stem colour Green/White/Yellow/Dark/Light/Other notes

Petiole length: base of plant cm top of plant cm

Rachis+petiole (without cirrus) total length: cm notes:

Leaflets: not grouped/grouped number in group

Number: on one side colour (top side) (under)

Other notes

Flagellum/Cirrus/Climbing organ not seen/Does not have climbing organ

Inflorescence length without flagellum cm

Partial Inflorescence: number PI cm apart PI cm long

Other notes

Fruit

Local names 1: 2:

What language? 1: 2:

Uses: Handicrafts Sell to traders Eat shoots Other:

Bold numbers show the main account for each species; references to that species in other species accounts are not included in this index. The letter of the relevant field key (pages 10-16) for each species is also given.

Subjects

Species